SPACE SCIENCE, EXPLORATION AND POLICIES

COMMERCIAL CREW PROGRAM, SPACE TELESCOPES AND THE GLOBAL POSITIONING SYSTEM

SPACE SCIENCE, EXPLORATION AND POLICIES

Additional books and e-books in this series can be found on Nova's website under the Series tab.

SPACE SCIENCE, EXPLORATION AND POLICIES

COMMERCIAL CREW PROGRAM, SPACE TELESCOPES AND THE GLOBAL POSITIONING SYSTEM

GRAEME N. LUTZ
EDITOR

snova
New York

Library of Congress Cataloging-in-Publication Data

ISBN: 978-1-53614-980-7

Published by Nova Science Publishers, Inc. † New York

CONTENTS

Preface		**vii**
Chapter 1	NASA Commercial Crew Program: Plan Needed to Ensure Uninterrupted Access to the International Space Station *United States Government Accountability Office*	**1**
Chapter 2	NASA Commercial Crew Program: Continued Delays Pose Risks for Uninterrupted Access to the International Space Station *Cristina T. Chaplain*	**53**
Chapter 3	Commercial Space Industry Launches a New Phase *Bill Canis*	**77**
Chapter 4	NASA: Preliminary Observations on the Management of Space Telescopes *Cristina T. Chaplain*	**107**
Chapter 5	James Webb Space Telescope: Integration and Test Challenges Have Delayed Launch and Threaten to Push Costs Over Cap *United States Government Accountability Office*	**129**

Chapter 6 Global Positioning System: Better Planning and Coordination Needed to Improve Prospects for Fielding Modernized Capability **163**
United States Government Accountability Office

Index **213**

PREFACE

NASA's Commercial Crew Program is a multi-phased effort that began in 2010. Across the phases, NASA has engaged several companies, using both agreements and contract vehicles to develop and demonstrate crew transportation capabilities. As the program has passed through these phases, NASA has generally narrowed down the number of participants. NASA's mission is to drive advances in science, technology, aeronautics, and space exploration, and contribute to education, innovation, our country's economic vitality, and the stewardship of the Earth. To accomplish this mission, NASA establishes programs and projects that rely on complex instruments and spacecraft. NASA's portfolio of major projects ranges from space satellites equipped with advanced sensors to study the Earth to a telescope intended to explore the universe to spacecraft to transport humans and cargo to and beyond low-Earth orbit.

Chapter 1 - In 2014, NASA awarded two firm-fixedprice contracts to Boeing and SpaceX, worth a combined total of up to $6.8 billion, to develop crew transportation systems and conduct initial missions to the ISS. In February 2017, GAO found that both contractors had made progress, but their schedules were under mounting pressure. The contractors were originally required to provide NASA all the evidence it needed to certify that their systems met its requirements by 2017. A House report accompanying H.R. 5393 included a provision for GAO to review the progress of NASA's human exploration programs. This chapter examines the Commercial Crew Program, including (1) the extent to which the contractors have made progress towards certification and (2) how NASA's certification process addresses safety of the contractors' crew transportation systems. GAO analyzed contracts, schedules, and other documentation and spoke with officials from NASA, the Commercial Crew Program, Boeing, SpaceX, and two of NASA's independent review bodies that provide oversight.

Chapter 2 - Since the Space Shuttle was retired in 2011, the United States has been relying on Russia to carry astronauts to and from the space station. NASA's Commercial Crew Program is facilitating private development of a domestic system to meet that need safely, reliably, and cost-effectively before the seats it has contracted for on a Russian spacecraft run out in 2019. In 2014, NASA awarded two firm-fixedprice contracts to Boeing and SpaceX worth a combined total of up to $6.8 billion to develop crew transportation systems and conduct initial missions to the space station. In February 2017, GAO found that both contractors had made progress, but their schedules were under mounting pressure. This statement provides preliminary observations on the extent to which the contractors and the program are making progress toward meeting NASA's standards for human spaceflight, a process called certification. This statement is based on ongoing work and information contained in GAO's February 2017 report on this program (GAO-17-137). To do this work, GAO analyzed contracts, schedules, and other documentation.

Chapter 3 - Rockets, satellites, and the services they provide, once the domain of governments, are increasingly launched and managed by privately owned companies. Although private aerospace firms have contracted with federal agencies since the onset of the Space Age six decades ago, U.S. government policy has sought to spur innovation and drive down costs by expanding the roles of satellite manufacturers and commercial launch providers. Global spending on space activity reached an estimated $323 billion in 2015. Of this amount, nearly 40% was generated by commercial space products and services and 37% by commercial infrastructure and support industries. The U.S. government—

including national security agencies and the National Aeronautics and Space Administration (NASA)—accounted for about 14% of global spending; government spending by other countries was responsible for the remaining 10%. The satellite and launch vehicle supply chains are global, with a small number of manufacturers. In 2015, global satellite manufacturing revenues were $6 billion; launches booked $2.6 billion in revenue. Ground stations—the largest part of the commercial space infrastructure—generated more than $100 billion in revenue, largely from geolocation and navigation equipment. The face of the U.S. space industry is changing with a government shift toward use of fixed price contracts for commercial services, new entrants with new launch products, and an increase in the use of smaller satellites:

- NASA's commercial cargo program and other federal contracts are supporting the growth of the commercial launch industry, with less expensive rockets, some of which are planned to be reusable. Many of the new space-related companies are attracting rising levels of venture capital.
- Aggressive pricing by U.S. entrants is cutting into the international launch market once dominated by foreign providers.
- A renewed interest in low-cost satellites, some of which are small enough to be held in one hand, is prompting a range of start-ups and providing new accessibility to space by educational institutions, small businesses, and individual researchers.

In order to spur innovation and growth, the commercial space industry has been purposely insulated from some types of federal regulation often applied to other industries. Nevertheless, three broad federal issues will affect the industry's future development. One is the structure of federal regulation and management; those responsibilities currently are dispersed among many agencies, and there is congressional interest in reorganizing commercial space functions at NASA and the Departments of Defense, Commerce, Transportation, and State. A second issue is the extent to which U.S. export controls are hampering U.S. satellite industry sales abroad. Export controls have recently been revamped to enable export of more commercial space products and services, but impediments may remain to reestablishing U.S. space product competitiveness. A third concern is that new Federal Communications Commission (FCC) regulations allowing wireless communication providers to share spectrum previously dedicated to satellite transmissions may result in interference. The commission has pledged to continue studying the issue.

Chapter 4 - Acquisition management has been a long-standing challenge at NASA, although GAO has reported on improvements the agency has made in recent years. Three space telescope projects are the key enablers for NASA to achieve its astrophysics' science goals, which include seeking to understand the universe. In its fiscal year 2018 budget request, NASA asked for about $697 million for these three projects, which represents over 50 percent of NASA's budget for its astrophysics' major projects. In total, these projects represent an expected investment of at least $12.4 billion. This statement reflects preliminary observations on (1) the current status and cost of NASA's major telescope projects and (2) lessons learned that can be applied to NASA's management of its telescope projects. This statement is based on ongoing work on JWST and ongoing work on the status of NASA's major projects. Both reports are planned to be published in Spring 2018. This statement is also based on past GAO reports on JWST and NASA's acquisitions of major projects, and NASA input.

Chapter 5 - JWST, a large, deployable telescope intended to be the successor to the Hubble Space Telescope, is one of NASA's most complex and expensive projects, at an anticipated cost of $8.8 billion. Congress set an $8 billion JWST development cost cap in 2011, and the remaining $837 million is for its operations costs. JWST is intended to revolutionize our understanding of star and planet formation and advance the search for the origins of our universe. With significant integration and testing planned for the remaining period until launch, the JWST project will still need to address many challenges during the remainder of integration and testing. Conference Report No. 112-284, accompanying the Consolidated and Further Continuing Appropriations Act, 2012, included a provision for

GAO to assess the project annually and report on its progress. This is the sixth such report. This chapter assesses the extent to which JWST is (1) meeting its schedule commitments, and (2) able to meet its cost commitments. GAO reviewed monthly JWST reports, reviewed relevant policies, conducted independent analysis of NASA and contractor data, and interviewed NASA and contractor officials.

Chapter 6 - GPS provides positioning, navigation, and timing data to civilian and military users who depend on this satellite-based system. Since 2000, DOD—led by the Air Force—has been working to modernize GPS and to keep the current system of satellites—known as the GPS constellation—operational, although these efforts have experienced cost and schedule growth. The National Defense Authorization Act for Fiscal Year 2016 contained a provision that the Air Force provide reports to GAO on GPS acquisition programs and that GAO brief the congressional defense committees. GAO briefed the committees in 2016 and 2017. This chapter summarizes and expands on information presented in those briefings. This chapter assesses the extent to which DOD faces acquisition challenges (1) sustaining the GPS constellation; (2) developing a new ground control system; and (3) developing and fielding modernized receivers. GAO analyzed GPS quarterly acquisition reports and data, acquisition strategies, software and test plans, and other documents, and interviewed DOD and contractor officials.

In: Commercial Crew Program …
Editor: Graeme N. Lutz
ISBN: 978-1-53614-980-7

Chapter 1

NASA COMMERCIAL CREW PROGRAM: PLAN NEEDED TO ENSURE UNINTERRUPTED ACCESS TO THE INTERNATIONAL SPACE STATION*

United States Government Accountability Office

ABBREVIATIONS

FAR	Federal Acquisition Regulation
ISS	International Space Station
NASA	National Aeronautics and Space Administration
ORDEM	Orbital Debris Engineering Model
SpaceX	Space Exploration Technologies Corporation

* This is an edited, reformatted and augmented version of a United States Government Accountability Office Report to Congressional Committees; Accessible Version, Publication No. GAO-18-476, dated July 2018.

WHY GAO DID THIS STUDY

In 2014, NASA awarded two firm-fixedprice contracts to Boeing and SpaceX, worth a combined total of up to $6.8 billion, to develop crew transportation systems and conduct initial missions to the ISS. In February 2017, GAO found that both contractors had made progress, but their schedules were under mounting pressure. The contractors were originally required to provide NASA all the evidence it needed to certify that their systems met its requirements by 2017.

A House report accompanying H.R. 5393 included a provision for GAO to review the progress of NASA's human exploration programs. This chapter examines the Commercial Crew Program, including (1) the extent to which the contractors have made progress towards certification and (2) how NASA's certification process addresses safety of the contractors' crew transportation systems. GAO analyzed contracts, schedules, and other documentation and spoke with officials from NASA, the Commercial Crew Program, Boeing, SpaceX, and two of NASA's independent review bodies that provide oversight.

WHAT GAO RECOMMENDS

GAO is making five recommendations, including that NASA develop a contingency plan for ensuring a U.S. presence on the ISS and clarify how it will determine its risk tolerance for loss of crew. NASA concurred with three recommendations; partially concurred on the recommendation related to loss of crew; and non-concurred with a recommendation to report its schedule analysis to Congress. GAO believes these recommendations remain valid, as discussed in the chapter.

WHAT GAO FOUND

Both of the Commercial Crew Program's contractors, Boeing and Space Exploration Technologies Corporation (SpaceX), are making progress finalizing designs and building hardware for their crew transportation systems, but both contractors continue to delay their certification milestone (see figure). Certification is the process that the National Aeronautics and Space Administration (NASA) will use to ensure that each contractor's system meets its requirements for human spaceflight for the Commercial Crew Program.

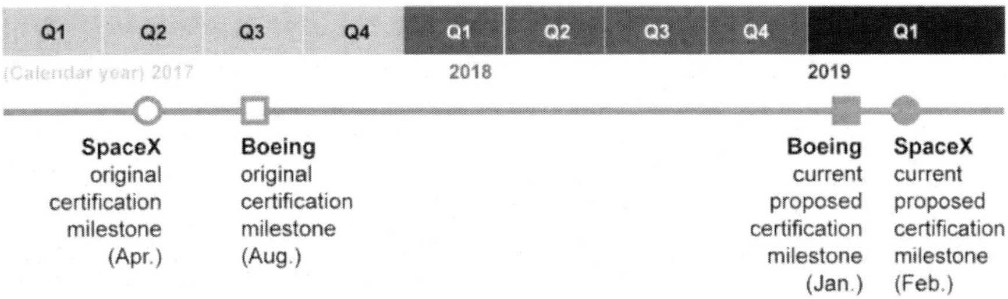

Source: GAO analysis of National Aeronautics and Space Administration contracts and documents. | GAO-18-476.

Commercial Crew Program: SpaceX and Boeing's Certification Delays as of First Quarter Calendar Year 2018.

Further delays are likely as the Commercial Crew Program's schedule risk analysis shows that the certification milestone is likely to slip. The analysis identifies a range for each contractor, with an earliest and latest possible completion date, as well as an average. The average certification date was December 2019 for Boeing and January 2020 for SpaceX, according to the program's April 2018 analysis. Since the Space Shuttle was retired in 2011, the United States has been relying on Russia to carry astronauts to and from the International Space Station (ISS). Additional delays could result in a gap in U.S. access to the space station as NASA has contracted for seats on the Russian Soyuz spacecraft only through November 2019. NASA is considering potential options, but it does not have a contingency plan for ensuring uninterrupted U.S. access.

NASA's certification process addresses the safety of the contractors' crew transportation systems through several mechanisms, but there are factors that complicate the process. One of these factors is the loss of crew metric that was put in place to capture the probability of death or permanent disability to an astronaut. NASA has not identified a consistent approach for how to assess loss of crew. As a result, officials across NASA have multiple ways of assessing the metric that may yield different results. Consequently, the risk tolerance level that NASA is accepting with loss of crew varies based upon which entity is presenting the results of its assessment. Federal internal controls state that management should define risk tolerances so they are clear and measurable. Without a consistent approach for assessing the metric, the agency as a whole may not clearly capture or document its risk tolerance with respect to loss of crew.

July 11, 2018

The Honorable Jerry Moran
Chairman

The Honorable Jeanne Shaheen
Ranking Member
Subcommittee on Commerce, Justice, Science, and Related Agencies
Committee on Appropriations
United States Senate

The Honorable John Culberson
Chairman

The Honorable José Serrano
Ranking Member
Subcommittee on Commerce, Justice, Science, and Related Agencies
Committee on Appropriations
House of Representatives

Following the retirement of the Space Shuttle in 2011, the United States was left with no domestic ability to provide crew access to the International Space Station (ISS). Since then, the National Aeronautics and Space Administration (NASA) has relied on obtaining seats on the Russian

Soyuz spacecraft to maintain a U.S. presence on the station. NASA's Commercial Crew Program is facilitating the commercial development of a crew transportation system that can provide safe, reliable, and cost-effective transportation to and from the ISS and that would end this dependency. NASA's goal is to have one or more contractors that can provide crew transportation services to the ISS. NASA is planning for the ISS to be operational until at least 2024.

NASA's acquisition strategy for the Commercial Crew Program is similar to the one it used on the Commercial Cargo Program, but different from other spacecraft it has built for humans, from Mercury to Gemini and Apollo to the Space Shuttle.[1] Each contractor is to design, develop, build, own, and operate its own spaceflight system and infrastructure. The contractors will have access to NASA's expertise and resources throughout the development process, but NASA engineers are not the ones making design decisions and NASA personnel will be less involved in processing, testing, launching, and operating the crew transportation system. In the end, NASA will buy a crew transportation service—a ride for its astronauts to and from the ISS—much like it does for ISS cargo.

In the most recent phase of the Commercial Crew Program, NASA awarded firm-fixed-price contracts in 2014 to Boeing and Space Exploration Technologies Corporation (SpaceX), valued at up to $4.2 billion and $2.6 billion, respectively, for the development of crew transportation systems that meet NASA requirements and for the initial service missions to the ISS. According to the contracts, the companies were originally supposed to complete certification—the process by which each contractor provides NASA the evidence it needs to certify that its systems meet performance and safety requirements—by 2017.

The House Report accompanying H.R. 5393, Commerce, Justice, Science, and Related Agencies Appropriations Bill, 2017 includes a

[1] In November 2005, NASA established the Commercial Crew and Cargo program office at the Johnson Space Center to challenge the commercial space industry to establish capabilities and services that could support the ISS's crew and cargo transportation needs. For more information, see GAO, *NASA: Commercial Partners Are Making Progress, but Face Aggressive Schedules to Demonstrate Critical Space Station Cargo Transport Capabilities,* GAO-09-618 (Washington, D.C.: June 16, 2009).

provision for GAO to review the acquisition progress of NASA's human exploration programs, including the Commercial Crew Program. For this review, we assessed (1) the extent to which the contractors have made progress towards meeting NASA's certification requirements and NASA's plans to ensure continued access to the ISS; and (2) how NASA's certification process addresses safety of the contractors' crew transportation systems. To assess the contractors' progress toward certification, we reviewed Commercial Crew Program and contract documents, including quarterly progress updates, monthly risk charts, and monthly schedule summaries. We also interviewed program and contract officials to identify steps being taken to mitigate risks and assess the extent of cost or schedule effects if the risk is realized. For each contractor, we compared the original contract schedule to the most current contract schedule and the contractors' internal development schedules to determine upcoming events and expected delays. We also spoke with ISS program officials to determine how NASA plans to mitigate the effects of these delays on its access to the ISS.

To assess how NASA's certification process addresses safety of the contractors' crew transportation systems, we reviewed agency safety policies, program plans, and contract documents to establish when certification approval would be granted, what safety assessments were required of the program, and which safety factors would be considered in certification reviews. For each contractor, we compared agency requirements to contract requirements for a key safety metric known as "loss of crew," and interviewed program and agency officials to determine how loss of crew would be assessed and considered throughout the certification process. We reviewed program and agency documentation and interviewed program and agency officials to determine the role of the safety and mission assurance technical authority in the program. We also met with two organizations that provide NASA with independent assessments of the program, the program's standing review board and the Aerospace Safety Advisory Panel, to gain their perspectives on the contractor's progress and how NASA addresses safety in its certification process. Appendix I contains detailed information about our scope and methodology.

We conducted this performance audit from April 2017 to July 2018 in accordance with generally accepted government auditing standards. Those standards require that we plan and perform the audit to obtain sufficient, appropriate evidence to provide a reasonable basis for our findings and conclusions based on our audit objectives. We believe that the evidence obtained provides a reasonable basis for our findings and conclusions based on our audit objectives.

BACKGROUND

NASA's Commercial Crew Program is a multi-phased effort that began in 2010. Across the phases, NASA has engaged several companies, using both agreements and contract vehicles to develop and demonstrate crew transportation capabilities. As the program has passed through these phases, NASA has generally narrowed down the number of participants. The early phases of the program were under Space Act agreements, which is what NASA calls the agreements entered into pursuant to its other transaction authority.[2] These types of agreements are generally not subject to the Federal Acquisition Regulation (FAR) and allow the government and its contractors greater flexibility in many areas. Under these Space Act agreements, NASA relied on the commercial companies to propose specifics related to their crew transportation systems, including their design, the capabilities they would provide, and the level of private investment. In these phases, NASA provided technical support and determined whether the contractors met certain technical milestones. In most cases, NASA also provided funding.

For the final two phases of the program, NASA awarded FAR-based contracts. By using FAR-based contracts, NASA gained the ability to procure missions to the ISS, while continuing to provide technical

---

[2] 51 U.S.C. § 20113(e). This authority allows NASA to enter into agreements "other than" standard government contracts or other traditional mechanisms.

expertise and funding to the contractors. NASA levied two sets of requirements on the contractors:

- the ISS program requirements, which must be met by all spacecraft visiting the ISS whether they carry cargo or crew; and
- the Commercial Crew Program requirements, which have a focus on system capabilities and safety rather than design.

The program also established a verification closure notice process, in which the contractors submit data to NASA to verify they have met all the requirements to be certified. This certification must occur before contractors are allowed to fly initial crewed missions to the ISS.

Current Program Contracts

In September 2014, NASA awarded firm-fixed-price contracts to Boeing and SpaceX, valued at up to $4.2 billion and $2.6 billion, respectively, for the Commercial Crew Transportation Capability phase. Under a firmfixed-price contract, the contractor must perform a specified amount of work for the price negotiated by the contractor and government. This is in contrast to a cost-reimbursement contract, in which the government generally agrees to pay the contractor's allowable costs regardless of whether work is completed. During this phase, the contractors will complete development of crew transportation systems.

Boeing's spacecraft—CST-100 Starliner—is composed of a crew module and a service module.

- The crew module will carry the crew and cargo. It also includes communication systems, docking mechanisms, and return systems for Earth landing.
- The service module provides propulsion on-orbit and in abort scenarios as well as radiators for thermal control.

SpaceX's spacecraft—Dragon 2—is composed of a capsule, which we refer to as the crew module, and a trunk, which we refer to as the support module.

- The crew module is composed of a pressure section and a service section. This module will carry the crew and cargo. It also includes avionics, docking mechanisms, and return systems for a water landing.
- The support module includes solar arrays for on-orbit power and guidance fins for escape abort scenarios.

Figure 1 shows the spacecraft and launch vehicles for Boeing and SpaceX's crew transportation systems.

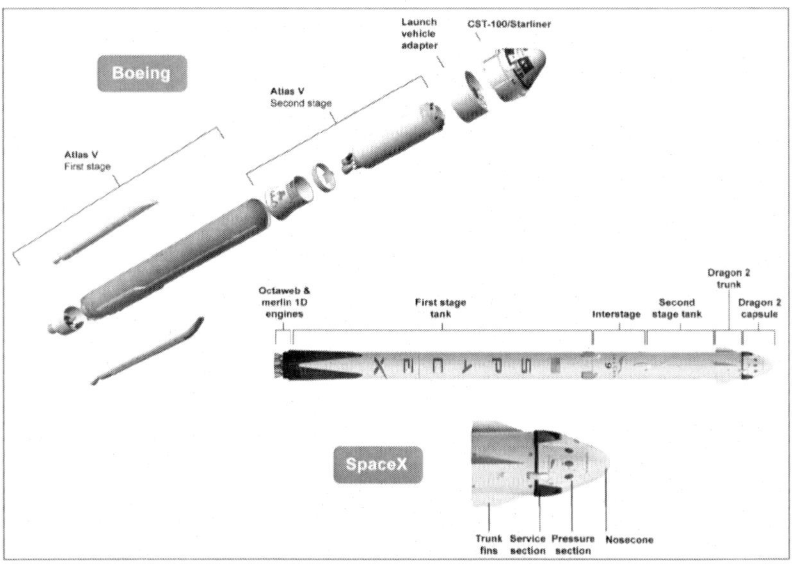

Source: Copyright © 2016 United Launch Alliance, LLC. (top image); @ 2016 SpaceX (bottom image). | GAO-18-476.

Figure 1. Boeing and SpaceX Crew Transportation Systems.

The Commercial Crew Transportation Capability phase contracts include three types of services:

- Contract Line Item 001 encompasses the firm-fixed-price design, development, test, and evaluation work needed to support NASA's certification of the contractor's spacecraft, launch vehicle, and ground support systems.
- Contract Line Item 002 covers any service missions that NASA orders to transport astronauts to and from the ISS. Under this indefinite-delivery, indefinite-quantity line item, NASA has ordered six post-certification missions from each contractor.[3] Each service mission is its own firm-fixed-price task order. NASA must certify the contractors' systems before they can fly these missions.
- Contract Line Item 003 is an indefinite-delivery, indefinite-quantity line item for any special studies, tests, and analyses that NASA may request. These tasks do not include any work necessary to accomplish the requirements under contract line item 001 and 002. As of April 2018, NASA had funded studies worth approximately $30 million to Boeing, including approximately $27 million for additional testing of the parachute system. NASA had funded studies worth approximately $44 million to SpaceX, including approximately $34 million for additional testing of the parachute system. For each contractor, the maximum value of this contract line item is $150 million.

NASA has made changes to the contracts that have increased their value. While the contracts are fixed-price, their values can increase if NASA adds work or otherwise changes requirements, among other means. As of April 2018, NASA requirement changes had increased the value of contract line item 001 for Boeing by approximately $191 million and for SpaceX by approximately $91 million.

---

[3] An indefinite-delivery, indefinite-quantity contract may be used to acquire supplies or services during a specified period when the exact times and exact quantities of future requirements are not known at the time of contract award.

Certification

NASA divided the certification work under contract line item 001 into two acceptance events: the design certification review and the certification milestone. An acceptance event occurs when NASA approves a contractor's designs and acknowledges that the contractor's work is complete and meets the requirements of the contract.

The first acceptance event—the design certification review—verifies the contractor's crew transportation system's capability to safely approach, dock, mate, and depart from the ISS, among other requirements. After the contractor has successfully completed all of its flight tests, as well as various other activities, the second acceptance event—the certification milestone—determines whether the crew transportation system meets the Commercial Crew Program's requirements. Following this contract milestone is an agency certification review, which authorizes the use of a contractor's system to transport NASA crew to and from the ISS. Figure 2 shows a notional path leading up to the agency certification review.

Source: GAO analysis of National Aeronautics and Space Administration (NASA) documents. | GAO-18-476.

Figure 2. Notional Commercial Crew Program Path to Certification by NASA.

The Commercial Crew Program's certification plan outlines how the program will incrementally review required deliverables leading up to, and supporting, the agency certification review. For each review, the plan describes the information that the contractor and the program will present. At the agency certification review, which is chaired by the Associate Administrator of the Human Exploration and Operations Mission Directorate, the agency will review the program's formal recommendation to certify the contractor's crew transportation system. Program officials said that their goal is to develop and review certification evidence incrementally in order to reduce the risk that issues will be identified during the agency certification review.

Prior GAO Work

In our February 2017 report,[4] we evaluated the progress made by the two contractors on the Commercial Crew Program and found the following:

- Both of the Commercial Crew Program's contractors had made progress developing their crew transportation systems, but both also had aggressive development schedules that were increasingly under pressure. We reported that both Boeing and SpaceX had determined that they would not be able to meet their original 2017 certification dates, and both expected certification to be delayed until 2018. We found that the schedule pressures were amplified by NASA's need to provide a viable crew transportation option because its contract with Russia's space agency was to provide crew transportation to the ISS for six astronauts through 2018 with rescue and return through late spring 2019. Purchasing additional seats from Russia involves a contracting process that typically takes 3 years. Without a viable contingency option for ensuring uninterrupted access to the ISS in the event of further Commercial Crew delays, we concluded that NASA was at risk of not being able to maximize the return on its multibillion dollar investment in the space station.
- The Commercial Crew Program was using mechanisms laid out in its contracts to gain a high level of visibility into the contractors' crew transportation systems, but maintaining that level of visibility through certification could add schedule pressures. We noted that, for example, due to NASA's acquisition strategy for this program,

[4] GAO, *NASA Commercial Crew Program: Schedule Pressure Increases as Contractors Delay Key Events,* GAO-17-137 (Washington, D.C.: Feb. 16, 2017).

its personnel were less involved in the testing, launching, and operation of the crew transportation system. While the program developed productive working relationships with both contractors, obtaining the level of visibility that the program required had also taken more time than the program or contractors had anticipated. Ultimately, we noted that the program had the responsibility for ensuring the safety of U.S. astronauts, and its contracts gave it deference to determine the level of visibility required to do so. We concluded that the program office could face difficult choices moving forward about how to maintain the level of visibility it feels it needs without adding to the program's schedule pressures.

In order to ensure that the United States had continued access to the ISS if the Commercial Crew Program's contractors experienced additional schedule delays, we recommended in our February 2017 report that the NASA Administrator develop a contingency plan for maintaining a presence on the ISS beyond 2018, including options to purchase additional Russian Soyuz seats, and report to Congress on the results.[5] NASA concurred with this recommendation, and in February 2017, NASA executed a contract modification that purchased two seats and included an option to purchase three additional crewmember seats from Boeing on the Russian Soyuz vehicle. These seats represent a contingency plan for U.S. access to the ISS through 2019. In April 2017, NASA informed the Congress of this action.

Contractors Have Made Progress, but NASA Has Not Finalized Plans to Ensure ISS Access Given Persistent Delays

Boeing and SpaceX continue to make progress developing their crew transportation systems, but both contractors have further delayed the

[5] GAO-17-137.

certification milestone to early 2019. These changes have occurred as the contractors continue to work to aggressive schedules, and they have had to delay key events regularly. Further delays are likely as the Commercial Crew Program's schedule risk analysis shows that the certification milestone is likely to further slip. In addition, as of mid-June 2018, NASA officials told us that these dates may change soon but that both contractors have not yet provided official updates to their schedules to NASA. NASA has not fully shared information with Congress regarding the risks of future schedule delays for the contractors and, as a result, Congress lacks insight into when the contractors will be certified. Also, there may be a gap in access to the ISS if the Commercial Crew Program experiences additional delays. While NASA has begun to discuss potential options, it currently does not have a contingency plan for how to ensure an uninterrupted presence on the ISS beyond 2019.

Contractors Continue to Make Progress, but Risks Remain

Boeing and SpaceX have continued to make progress finalizing their designs and building hardware as they work toward their certification milestones. The contractors are manufacturing test articles to demonstrate system performance and flight spacecraft to support the uncrewed and crewed flight tests, which are expected to demonstrate the ability to meet contract requirements. As table 1 shows, these test articles and spacecraft vary in levels of completion. Some are built and undergoing testing while others are starting the manufacturing phase. Should any issues arise during integration and test or the flight tests, the contractors may have to complete rework on the spacecraft already under construction.

While both contractors are making progress, the Commercial Crew Program is tracking risks that each contractor has to address through testing and other means as they work towards the certification milestone.

Table 1. Description of Boeing and SpaceX Hardware, Current Status, and Upcoming Events as of Second Quarter Calendar Year 2018

Spacecraft (Name and Type)	Purpose	Current Status	Upcoming Events
Boeing			
Flight spacecraft 1	Environmental testing Crewed flight test Second post-certification mission	Crew module: constructed, in final outfitting Service module: constructed, in final outfitting	Boeing plans to conduct environmental testing starting in the third quarter of 2018 to test the spacecraft in conditions that simulate the space environment.
Flight spacecraft 2	Uncrewed flight test First post-certification mission	Crew module: constructed, in initial outfitting Service module: in construction	Boeing plans to join the crew and service modules together in the third quarter of 2018.
Test article 1	Validate effectiveness of spacecraft design and abort system	Crew module: constructed and integrated Service module: constructed and integrated	This test article has undergone testing throughout 2017 and is expected to complete testing in mid-2018.
Test article 2	Support ground tests Padabort test	Crew module: constructed and integrated Service module: in construction	This test article is completing testing before it will be reconfiguredto support the pad abort test.
SpaceX			
Flight spacecraft 1	Uncrewed flight test	Crew module: constructed and integrated Support module: in construction	SpaceX plans to join the crew and support modules together in the second quarter of 2018.
Flight spacecraft 2	Crewed flight test	Crew module: in construction Support module: in construction	SpaceX plans to join the crew and support modules together in the third quarter of 2018.
Flight spacecraft 3	First post-certification mission	Crew module: in construction Support module: in construction	SpaceX plans to join the crew and support modules together in the first quarter of 2019.
Test article	Support spacecraft propulsion testing	Testing is underway to validate performance of the spacecraft engine propulsion system.	SpaceX plans to complete this testing by the third quarter of 2018.

Source: GAO analysis of National Aeronautics and Space Administration and contractor documents. | GAO-18-476.

As we have previously reported, these types of risks are inherent in NASA's major acquisitions, which are highly complex, specialized, and often pushing the state of the art in space technology, but they could also delay the contractors' progress if issues arise during testing.[6]

The Commercial Crew Program's top programmatic risks identified for Boeing include challenges related to its abort system performance, parachutes, and launch vehicle.

- Abort System: Boeing is addressing a risk that its abort system, which it needs for human spaceflight certification, may not meet the program's requirement to have sufficient control of the vehicle through an abort. In some abort scenarios, Boeing has found that the spacecraft may tumble, which could pose a threat to the crew's safety. To validate the effectiveness of its abort system, Boeing has conducted extensive wind tunnel testing and plans to complete a pad abort test in July 2018.
- Parachute System: Boeing is also addressing a risk that during descent, a portion of the spacecraft's forward heat shield may re-contact the spacecraft after it is jettisoned and damage the parachute system.[7] Boeing's analysis indicates the risk exists only if one of two parachutes that pull the forward heat shield away from the spacecraft does not deploy as expected, and that potential re-contact is non-detrimental. However, NASA's independent analysis indicates that this may occur even if both parachutes deploy as expected. If the program determines this risk is unacceptable, Boeing would need to redesign the parachute system, which the program estimates could result in at least a 6-month delay.
- Launch Vehicle Data: One of the program's top programmatic and safety concerns is that it may not have enough information from Boeing's launch vehicle provider, United Launch Alliance, to

[6] GAO-17-137.

[7] The forward heat shield protects the parachute system during re-entry to the Earth's atmosphere.

assess whether the Atlas V launch vehicle prevents or controls cracking that could lead to catastrophic failures. NASA estimates that unfinished work in this area could take Boeing and the United Launch Alliance until the fourth quarter of 2018 to complete. Additionally, the first stage of the Atlas V is powered by the Russian built RD-180 engine, and, according to program and Boeing officials, access to its data is highly restricted by agreements between the U.S. and Russian governments. Since our last report, the Commercial Crew Program has lowered the risk that certification of the launch vehicle might not occur by negotiating steps to access necessary data, but work is still ongoing.[8]

The Commercial Crew Program's top programmatic risks identified for SpaceX are in part related to ongoing design and development efforts related to its launch vehicle design, the Falcon 9 Block 5.

- Composite Overwrap Pressure Vessel: This Block 5 design includes SpaceX's redesign of the composite overwrap pressure vessel, which is intended to contain a gas under high pressure. SpaceX officials stated the newly designed vessel aims to eliminate risks identified in the older design, which was involved in an anomaly that caused a mishap in September 2016. SpaceX plans to qualify the updated design for flight prior to the uncrewed flight test design certification review.
- Engine Turbine Cracking: The Block 5 design also includes design changes to address cracks in the turbine of its engine identified during development testing. NASA program officials told us that they had informed SpaceX that the cracks were an unacceptable risk for human spaceflight. SpaceX officials told us that they have made design changes to this Block 5 upgrade that did not result in any cracking during initial testing. However, this risk will not be

[8] GAO-17-137.

closed until SpaceX successfully completes qualification testing in accordance with NASA's standards without any cracks. As of March 2018, SpaceX had not yet completed this testing.

- Propellant Loading Procedures: Both the program and a NASA advisory group have raised SpaceX's plan to fuel the launch vehicle after the astronauts are on board the spacecraft to be a potential safety risk. In the May 2018 meeting minutes, however, the Aerospace Safety Advisory Panel stated that with appropriate controls in place, this approach could be a viable option for the program to consider.[9] SpaceX's perspective is that this operation may be a lower risk to the crew because it reduces the crew exposure time while the launch vehicle is being loaded with propellant. To better understand the propellant loading procedures, the program and SpaceX agreed to demonstrate the loading process five times from the launch site in the final crew configuration prior to the crewed flight test. The five events include the uncrewed flight test and the in-flight abort test. Therefore, delays to those events would lead to delays to the agreed upon demonstrations, which could in turn delay the crewed flight test and certification milestone.

Program's Schedule Risk Analysis Indicates More Delays Likely to Certification Milestone

Both contractors have notified NASA that their certification milestones have slipped to January 2019 for Boeing and February 2019 for SpaceX, but the Commercial Crew Program's schedule risk analysis indicates more delays are likely. This analysis identifies a range for each contractor, with an earliest and latest possible completion date, as well as an average. In April 2018, the program's schedule risk analysis found there was zero percent chance that either contractor would achieve its current proposed

[9] The Aerospace Safety Advisory Panel is an independent review body that provides advice and makes recommendations to the NASA Administrator on matters related to safety.

certification milestone. The analysis's average certification date was December 2019 for Boeing and January 2020 for SpaceX. Figure 3 shows the original Boeing and SpaceX contract schedules and the current proposed schedule for five key events in each contract, as well as NASA's schedule risk analysis for the certification milestone.

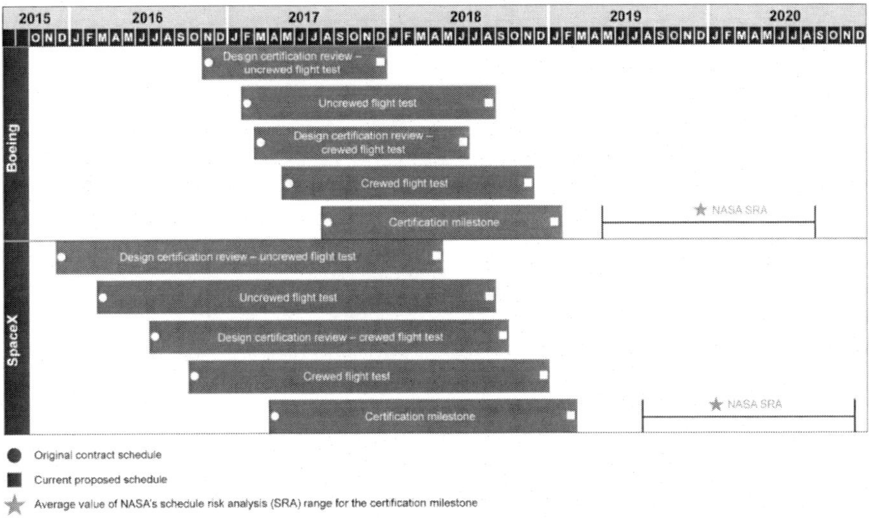

Source: GAO analysis of National Aeronautics and Space Administration (NASA) and contractor documents. | GAO-18-476.

Notes: The Commercial Crew Program's schedule risk analysis identifies a range, with an earliest and latest possible completion date, as well as an average.

The uncrewed and crewed flight tests are not contract milestones for Boeing. The dates for these events are from other Boeing or Commercial Crew documentation.

Figure 3. Boeing and SpaceX's Proposed Commercial Crew Schedule Delays, as of First Quarter Calendar Year 2018, and the Commercial Crew Program's April 2018 Schedule Risk Analysis Range.

Each month, the program updates its schedule risk analysis based on the contractors' internal schedules as well as program officials' perspectives and insight into specific technical risks. The Commercial Crew Program manager told us that differences between the contractors' proposed schedules and the program's schedule risk analysis include:

- The contractors are aggressive and use their schedule dates to motivate their teams, while NASA adds additional schedule margin for testing.
- Both contractors assume an efficiency factor in getting to the crewed flight test that NASA does not factor into its analysis.

The program manager also told us that the program meets with each contractor monthly to discuss schedules and everyone agrees to the relationships between events in the schedule even if they disagree on the length of time required to complete events. The program manager added, however, that she relies on her prior experience to estimate schedule time frames as opposed to relying on the contractors' schedules, which are often optimistic.

Our analysis also shows that the contractors often delay their schedules. Both contractors have repeatedly stated that their schedules are aggressive and have set ambitious—rather than realistic—dates, only to frequently delay them. Since the current contracts were awarded in 2014, the Commercial Crew Program has held 13 quarterly reviews for each contractor. For the five key events identified above, Boeing has reported a delay at 7 of those quarterly reviews and SpaceX has reported a delay at 9 of them.[10]

In mid-June 2018, NASA officials told us that the dates for these key events may change soon. The information presented in Figure 3 above is based on first quarter calendar year 2018 data. NASA officials stated both contractors have not yet officially communicated new schedule dates to NASA as of the second quarter calendar year 2018. We found that both contractors have updated schedules that indicate delays are forthcoming for at least one key event, but NASA officials told us they lack confidence in those dates until they are officially communicated to NASA by the contractors. As a result, NASA is managing a multibillion dollar program

[10] At SpaceX's first quarter review in 2018, the contractor identified at least one key event was being reviewed—the design certification review for the uncrewed flight test—but this had not officially been delayed so we did not count that as a delay in our analysis.

without confidence in its schedule information as it approaches several big events, including uncrewed and crewed flight tests.

NASA Has Neither Shared Complete Information on Delay Risk with Congress nor Developed a Contingency Plan

The risk of future delays in the contractors' schedules is critical information that NASA has not fully shared with Congress. Moreover, NASA has not yet developed a contingency plan to address the potential gaps that these delays could have on U.S. access to the ISS after 2019.

Specifically, in the Explanatory Statement accompanying the fiscal year 2018 Consolidated and Further Continuing Appropriations Act, the House Appropriations Committee stated its expectation that NASA report quarterly to the Senate and House Committees on Appropriations on the status of the Commercial Crew Program contracts.[11] Previously, members of Congress had asked for this information in order to ensure that Congress had adequate insight into this program. While NASA includes both contractors' proposed schedules in its quarterly report to Congress, NASA does not include the results of its own schedule risk analysis. Given the frequency with which the contractors delay key events in their schedules, the program's schedule risk assessment provides valuable insight into potential delays that NASA currently is not providing to Congress.

In addition, as previously mentioned, NASA executed a contract modification that purchased two seats and included an option to purchase three additional crew member seats through Boeing for an undisclosed value and reported this action to Congress in April 2017.[12] Ultimately, the option was exercised, and NASA purchased a total of five seats on four different Soyuz flights. Boeing obtained these seats through a separate settlement with the Russian firm RSC Energia, which manufactures the Soyuz.

---

[11] H.R. Rep. No. 115-231.

[12] In 2015, NASA paid approximately $82 million per seat through its contract with the Russian Federal Space Agency (Roscosmos). See GAO-17-137.

These seats were intended to serve as a contingency plan based on schedule information available at that time. However, subsequent delays, as well as the risk of future delays as discussed above, indicate that this contingency plan will likely no longer be sufficient. The earliest and latest possible completion dates for certification in NASA's April 2018 schedule risk analysis indicate it is possible that neither contractor would be ready before August 2020, leaving a potential gap in access of at least 9 months. We calculated the potential gap based on the contractor certification milestone dates, but there could be some additional time required between that review and the first post-certification service mission to the ISS.

As seen in figure 4, if the contractors can maintain their current proposed schedules for their respective certification milestones, a gap in access to the ISS is not expected. However, there would be a gap in access to the ISS if neither contractor has its certification milestone before November 2019, which is when NASA expects the final Russian Soyuz seat for a U.S. astronaut to return.

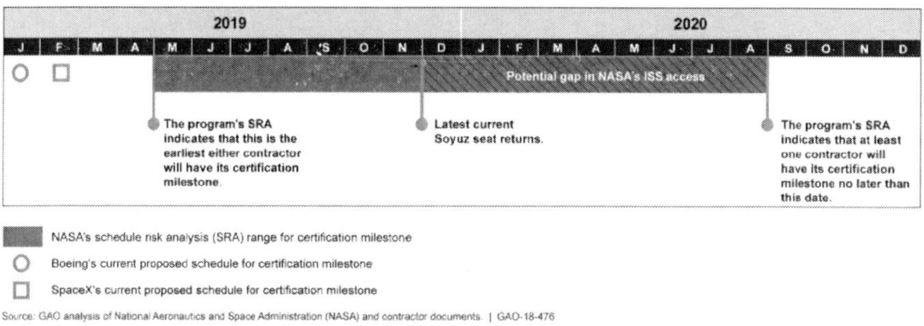

Source: GAO analysis of National Aeronautics and Space Administration (NASA) and contractor documents. | GAO-18-476.

Figure 4. Potential Gap in Access to the International Space Station (ISS) Based on the Commercial Crew Program's Schedule Risk Analysis Range as of April 2018.

Senior NASA officials told us that sustaining a U.S. presence on the ISS is essential to maintain and operate integral systems, without which the ISS cannot function. Given the importance of maintaining a U.S. presence on the ISS, NASA officials have stated they are working on options to address the potential gap in access. However, officials told us that planning for contingencies is difficult given the extensive international negotiations required for some options. Obtaining additional Soyuz seats seems unlikely, as the process for manufacturing the spacecraft and contracting for those seats typically takes 3 years—meaning additional seats would not be available before 2021. As a result, according to NASA's Associate Administrator for Human Exploration and Operations, the options NASA is considering include:

- Refine the remaining Soyuz launch schedule to allow for a return in January 2020, as opposed to November 2019. This would provide 2 additional months of access to the ISS before the commercial crew flights need to start.
- Use the crewed flight tests as operational flights to transport U.S. astronauts to and from the ISS. In March 2018, NASA modified Boeing's contract to allow NASA to add a third crew member and extend the length of the flight test, if NASA chooses to do so. This would have limited usefulness, however, in filling a potential gap in access to the ISS if the schedule for Boeing's crewed flight test slips past the return date for the last Soyuz flight and SpaceX also continues to experience delays.

NASA's Associate Administrator for Human Exploration and Operations stated that he is "brainstorming" other options to ensure access to the ISS but does not have a formal plan. While options are not unlimited and decisions have to be made within the context of the current geopolitical environment, Congress stated in the NASA Authorization Act of 2005 that it is U.S. policy to possess the capability for human access to space on a continuous basis. In 2010, Congress further stated that one of the key objectives of the United States' human spaceflight policy is to sustain the capability for long-duration presence in low-Earth orbit through full utilization of the ISS. If NASA does not develop options for ensuring access to the ISS in the event of further Commercial Crew delays, it will not be able to ensure that the U.S. policy goal and objective for the ISS will be met.

Agency Certification Process Includes Mechanisms to Assess Safety, but Is Complicated by Assessment of Key Safety Metric and Oversight Structure

The Commercial Crew Program relies on several contractual mechanisms to assess safety throughout the certification process, and those mechanisms are in varying stages of completion. The program itself, its contractors, and two of NASA's independent review organizations have raised concerns about the program's ability to assess and evaluate all of the deliverables in a timely manner. In addition, one of the key safety requirements levied by the program is loss of crew, which captures the probability of death or disability to a crew member. NASA does not have a consistent approach for how to incorporate key inputs to assess this metric, which means the agency as a whole may not clearly capture or document its risk tolerance with respect to loss of crew. Further, the program's chief safety and mission assurance officer is dual hatted to serve simultaneously in a programmatic position as well as the program's safety technical authority. This approach creates an environment of competing interests because it relies on the same individual to manage technical and safety aspects on behalf of the program while also serving as the independent oversight of those same areas.

Program Has Several Contractual Mechanisms to Assess Safety

The contractors are required to provide several key deliverables to the Commercial Crew Program, which inform the agency certification review and help NASA determine the level of risk it is accepting with respect to safety of each spacecraft. As described below, these deliverables are in varying stages of completion and the program itself, its contractors, and two of NASA's independent review organizations have raised concerns about the program's ability to assess and evaluate all of the deliverables in a timely manner.

Certification Data Package

Among other things, the certification data package includes a list of seven system safety assessments. For example, the certification data package includes a fault tolerance assessment, which describes the system's ability to sustain a certain number of undesired events, such as software or operational anomalies. A human error analysis—one of the seven assessments in the data package—evaluates human errors to minimize their negative effects on the system.

- Boeing held its uncrewed flight test design certification review in December 2017 and submitted its certification data package for NASA approval. Boeing plans three more updates to this data package prior to the final certification milestone.
- SpaceX has begun to submit data and plans to submit its final certification data package as part of its crewed flight test design certification review, which is scheduled for September 2018.

According to the program's certification review plan, program officials will review and approve the contractors' certification data packages, which will be used to inform the agency certification review.

Phased Safety Review Process

A three-phased safety review process informs the program's quality assurance activities, and it is intended to ensure that the contractors have identified all safety-critical hazards and implemented associated controls prior to the first crewed flight test.

- In phase one, the contractors identified risks in their designs and developed reports on potential hazards, the controls they put in place to mitigate them, and explanations for how the controls will mitigate the hazards.

- In phase two, which is nearing completion, the program reviews and approves the contractors' hazard reports and develops strategies to verify and validate that the controls are effective. For example, if a control requires that an item be waterproofed, verification and validation strategies could include inspections and tests to confirm that the item is waterproof. As of April 2018, the program had yet to complete this phase,[13] having approved 97 percent of Boeing's phase two reports and 72 percent of SpaceX's phase two reports.
- In phase three, the contractors will conduct the verification activities and submit the hazard reports to the program for approval. The program has begun phase three, including approving 19 percent of Boeing's phase three reports.

Program Requirements

While the program manager told us that all of the requirements contribute to the safety of the commercial systems, safety officials are required to approve a subset of these requirements.

Table 2. Verification Closure Notice Progress for Commercial Crew Program Requirements, by Contractor as of March 2018

Contractor	Notices	Total	Number Approved	Percentage Approved
Boeing	Verification closure notices	275	30	11
	Safety-specific notices	89	2	2
SpaceX	Verification closure notices	278	7	3
	Safety-specific notices	89	0	0

Source: GAO analysis of National Aeronautics and Space Administration data. | GAO-18-476.

Notes: The contractors have also started to submit notices for the International Space Station program requirements for the program to approve prior to the certification milestone.

Safety–specific notices are a subset of verification closure notices that require safety officials' approval.

---

[13] In February 2017, we found that the program had planned to complete phase two in early 2016, but was behind schedule. See GAO-17-137.

Examples of requirements approved by safety officials include the ability to leave the spacecraft in an emergency or to abort a launch. When a contractor is ready for NASA to verify that it has met a requirement, the contractor submits data for NASA to review through a verification closure notice. We define "safety-specific notices" as those requiring safety officials' approval. As shown in table 2, as of March 2018, the program had approved 2 percent of Boeing's safety-specific notices and 0 percent of SpaceX's safety-specific notices.

Testing

The program also requires testing to verify and validate the crew transportation system. Agency officials emphasized the importance of testing to safety, stating that testing reduces uncertainty about a system's performance and can uncover unknown problems. As noted above, both contractors will be conducting an uncrewed and a crewed flight test prior to being certified. While a certain level of risk needs to be accepted to conduct human spaceflight, these flight tests help to mitigate this risk by validating the integrated performance of the hardware and software. Agency and program officials stated that the contractors' flight tests are critical evidence to support certification of a safe and reliable system.

As evidenced by the data above, the program still has a significant amount of work ahead with respect to approving certification packages and closing hazard reports and verification closure notices. We have previously found that the program's workload was an emerging schedule risk, and the contractors have continued to express concern about program officials' ability to process and approve certification paperwork in a timely manner.[14] Workload has also been a concern for two of NASA's independent review organizations. For example, the Aerospace Safety Advisory Panel noted in its January 2018 annual report that the sheer volume of work that remains for the program in terms of closing hazard reports and verification closure notices is significant. In addition, the program's safety and mission assurance office identified the upcoming

[14] GAO-17-137 and GAO-18-317T.

bow wave of work in a shrinking time period as a top risk to achieving certification.

NASA Lacks a Consistent Approach to Assess Key Safety Metric

One mechanism the program put in place to assess the overall safety of each spacecraft—loss of crew—has been a focus of the Aerospace Safety Advisory Panel, Members of Congress, our prior work, and the program itself. Loss of crew captures the probability of death or permanent disability to one or more crew members. It has received a lot of attention, in part, because it has been a top risk for the program since 2015. Specifically, the program has been concerned that neither contractor would be able to meet the contract requirement of a 1 in 270 probability of incurring loss of crew. We identified two key concerns with how NASA is using the loss of crew metric: (1) inconsistent approaches to assess the loss of crew metric and (2) no identified plan to share lessons learned about using the loss of crew metric as a safety threshold.

A loss of crew value is generated through a probabilistic safety analysis, which models scenarios that could result in the loss of crew using various inputs. According to the program's analysis, the probability of on-orbit debris damaging the vehicle has the greatest effect on a loss of crew value. This probability is informed by an orbital debris (debris) model, which was updated in 2014, after the loss of crew requirement was established. The updated debris model makes it harder to meet a loss of crew value, in part, because the modeling environment where the contractors' systems will operate has changed. For example, the updated model includes a larger span of orbit, greater range of debris sizes, and the addition of material density classifications, which were not included in the former model. Further, the probabilistic safety analysis may include operational mitigations, such as on-orbit inspections that would include using cameras on the ISS to visually survey the spacecraft for damage, which, according to officials, makes it easier to meet a loss of crew value.

NASA describes the probabilistic safety analysis as a powerful tool that should be used as part of the overall risk management process to ensure the risk associated with development and operation of a system is understood, evaluated, managed, and mitigated. However, we found differences in the approaches that officials plan to use to assess loss of crew as well as in the loss of crew value being measured that could limit the usefulness of this tool.

- Agency Certification. The agency certification review for each contractor will include an assessment of whether its crew transportation system meets a loss of crew threshold of 1 in 150 for missions to the ISS, which is based on a May 2011 safety memo from the Office of Safety and Mission Assurance. A loss of crew value with a higher denominator, such as 1 in 270, is harder to meet than with a lower denominator, such as 1 in 150. According to the Chief of the Office of Safety and Mission Assurance, he will assess the 1 in 150 threshold using a probabilistic safety analysis that includes the updated debris model and operational mitigations, such as the on-orbit inspections cited above.
- Program Office. According to program officials, they will assess whether either contractor meets a 1 in 270 loss of crew value based on a probabilistic safety analysis using the former debris model (not the updated model) and not including operational mitigations.
- Contracting Officer. According to the contracting officer, each contractor's loss of crew requirement is 1 in 270 without including operational mitigations. The contracting officer stated that SpaceX's contract requirement uses the updated debris model in the probabilistic safety analysis, whereas Boeing's contract requirement uses the former debris model in the probabilistic safety analysis.
- Program's Chief Safety and Mission Assurance Officer. According to the program's chief safety and mission assurance officer, he will conduct a probabilistic safety analysis using the updated debris model and will not include operational mitigations to assess

whether each contractor meets a 1 in 200 loss of crew value. This loss of crew value stems from a program update that occurred after the initial contracts were signed.

These different approaches are summarized in table 3 below.

Table 3. Loss of Crew Values and Probabilistic Safety Analysis Inputs NASA Plans to Include Vary across the Agency

Loss of Crew Value			
Agency Certification	**Program Office**	**Contracting Officer**	**Program Chief Safety and Mission Assurance Officer**
1 in 150	1 in 270	1 in 270	1 in 200
Orbital Debris Model		Operational Mitigations	
Updated		Included	
Former		Excluded	
Updated –SpaceX Former –Boeing		Excluded	
Updated		Excluded	

Source: GAO analysis of National Aeronautics and Space Administration data. | GAO-18-476.

Agency policy requires human spaceflight programs to set a safety threshold, which NASA did for the Commercial Crew Program when it identified the 1 in 150 loss of crew threshold in the May 2011 safety memo. Subsequently, the program set more rigorous loss of crew values in contract and program documents. NASA also updated the debris model, which we previously noted makes it more difficult to meet a loss of crew value. As a result, NASA does not have a consistent approach for how to incorporate key inputs to the probabilistic safety analysis, including changes to the debris model. Instead, the risk tolerance that NASA is accepting with loss of crew varies based upon which entity is presenting the results of its probabilistic safety analysis. For example, it is possible that the program's assessment will determine that neither contractor will meet the 1 in 270 contract requirement, but that the agency's assessment will determine that the contractors meet the 1 in 150 agency certification value because that analysis will include operational mitigations. Or the program's assessment could determine that Boeing meets the 1 in 270 contractual loss of crew requirement, but the agency's assessment may determine that Boeing does not meet the 1 in 150 agency certification value because that analysis will use the updated debris model.

Federal internal controls state that agency management should define objectives clearly to enable the identification of risks and define risk tolerances. Specifically, management should define risk tolerances in specific and measurable terms, so they are clearly stated and can be measured.[15] In this case, because there will be multiple analyses conducted using different inputs, NASA risks not clearly capturing or documenting, in a coherent manner, its overall risk tolerance with respect to loss of crew before a final decision must be made on whether to certify either crew transportation system.

Moreover, capturing the challenges and lessons learned from using the loss of crew metric is critical, particularly because agency officials told us that this is the first time this metric has been used as a safety threshold. Also, there are different viewpoints about the utility of the metric as a safety threshold across the agency. The program manager repeatedly told us that loss of crew is best used as a design tool. For example, program officials told us that both contractors incorporated additional orbital debris shielding into their designs to mitigate the orbital debris risk and improve their loss of crew values. In addition, the Aerospace Safety Advisory Panel reported in 2018 that loss of crew should not be viewed as an absolute measure of actual risk during operations.[16] However, the May 2011 agency safety memo states that a breach of the loss of crew threshold would initiate a termination review of the Commercial Crew Program, which is a more strict application of the loss of crew metric.

Both program and safety officials told us that, after the agency certification is complete and lessons learned are available to be compiled, sharing those lessons learned across NASA would be a good idea given the complexities associated with assessing the loss of crew metric. As of April 2018, however, agency officials said they did not have a plan for loss of

---

[15] GAO, *Standards for Internal Control in the Federal Government,* GAO-14-704G (Washington, D.C.: September 2014).

[16] Aerospace Safety Advisory Panel, *Annual Report for 2017* (Washington, D.C.: January 2018).

crew knowledge-sharing. We have previously found that lessons learned provide a powerful method of sharing good ideas for improving work processes, facility or equipment design and operation, quality, safety, and cost-effectiveness.[17] Further, according to NASA's Knowledge Policy on Program and Projects, which is managed through the Office of the Chief Engineer, a principle of each center and mission directorate's knowledge strategy is that knowledge is the cornerstone of NASA's ability to achieve mission success.[18] The policy acknowledges that NASA faces continuous challenges in using what it knows effectively. These challenges include, but are not limited to, enabling the identification and flow of knowledge across organizational boundaries; preserving knowledge at risk of being lost; and providing means for individuals, teams, and the organization to learn from experiences. If NASA does not capture lessons learned from the Commercial Crew Program on using the loss of crew requirement to set a program's safety threshold and whether it met the agency's intended goal, future programs will not be able to benefit from the knowledge gained from this multibillion dollar investment.

The Commercial Crew Program's Organizational Structure Impairs Independence of Safety Technical Oversight Supporting the Certification Process

NASA's governance model prescribes a management structure that employs checks and balances among key organizations to ensure that decisions have the benefit of different points of view and are not made in isolation. As part of this structure, NASA established the technical authority process as a system of checks and balances to provide independent oversight of programs and projects in support of safety and mission success through the selection of specific individuals with

---

[17] GAO, *NASA: Better Mechanisms Needed for Sharing Lessons Learned*, GAO-02-195 (Washington, D.C.: Jan. 30, 2002).

[18] NASA Policy Directive 7120.6, Knowledge Policy on Programs and Projects, para. 5.d(4)(a) (Nov. 26, 2013).

delegated levels of authority. The technical authority process has been used in other parts of the government for acquisitions, including the Department of Defense and Department of Homeland Security.

The Commercial Crew Program is organizationally connected to three technical authorities within NASA: the Office of the Chief Engineer technical authority, the Office of Chief Health and Medical technical authority, and the Office of Safety and Mission Assurance (safety) technical authority. The safety technical authority is responsible for ensuring from an independent standpoint that the program's products and processes satisfy NASA's safety, reliability, and mission assurance policies. The NASA safety technical authority has delegated authority through the Kennedy Space Center Director to the Chief Safety and Mission Assurance Officer for the Commercial Crew Program.

We have previously reviewed how NASA has organized its technical authorities for its Exploration Systems Development organization—an organization that oversees the development of the Space Launch System, Orion crew capsule, and associated ground systems that have the goal of extending human presence beyond low-Earth orbit. In October 2017, we found that the Exploration Systems Development organization had established an organizational structure in which the technical authorities for engineering and safety and mission assurance were dual hatted simultaneously in programmatic positions.[19] We found that having the same individual simultaneously fill both a technical authority role and a program role created an environment of competing interests, where the technical authority's ability to impartially and objectively assess the programs while at the same time acting on behalf of the Exploration Systems Development organization in programmatic capacities may be subject to impairments.

We found that this was in contrast to a recommendation from the Columbia Accident Investigation Board report—the result of an in-depth

---

[19] GAO, *NASA Human Space Exploration: Integration Approach Presents Challenges to Oversight and Independence*, GAO-18-28 (Washington, D.C.: Oct. 19, 2017).

assessment of the technical and organizational causes of the 2003 Space Shuttle Columbia accident—for NASA to establish a technical authority to serve independently of the Space Shuttle program, so that employees would not feel hampered to bring forward safety concerns or disagreements with programmatic decisions.[20] The board's findings that led to this recommendation included a broken safety culture in which it was difficult for minority and dissenting opinions to percolate up through the hierarchy; dual center and programmatic roles vested in one person that had confused lines of authority, responsibility, and accountability and made the oversight process susceptible to conflicts of interest; and oversight personnel in positions within the program, increasing the risk that these staffs' perspectives would be hindered by too much familiarity with the programs they were overseeing. In October 2017, we recommended that the division no longer dual hat two individuals who had both programmatic and technical authority responsibilities.[21] As of April 2018, NASA had taken steps to separate the engineering technical authority position from the programmatic position, and NASA's Chief of Safety and Mission Assurance said he planned to separate the safety position but had not yet completed that action.

The Commercial Crew Program employs a similar structure to the Exploration Systems Development organization in that the safety technical authority is dual hatted simultaneously in a programmatic position as the Commercial Crew Program's Safety and Mission Assurance Manager. According to the program's safety technical authority, in his programmatic role for the program, he helps set priorities for safety issues, including how staff will be utilized to meet those priorities. In the technical authority role, he provides independent oversight in support of safety and mission success. In his dual-hatted role, this official will be responsible for endorsing the program's certification recommendations in two different capacities: as the technical authority and as a program authority.

---

[20] GAO-18-28.
[21] GAO-18-28.

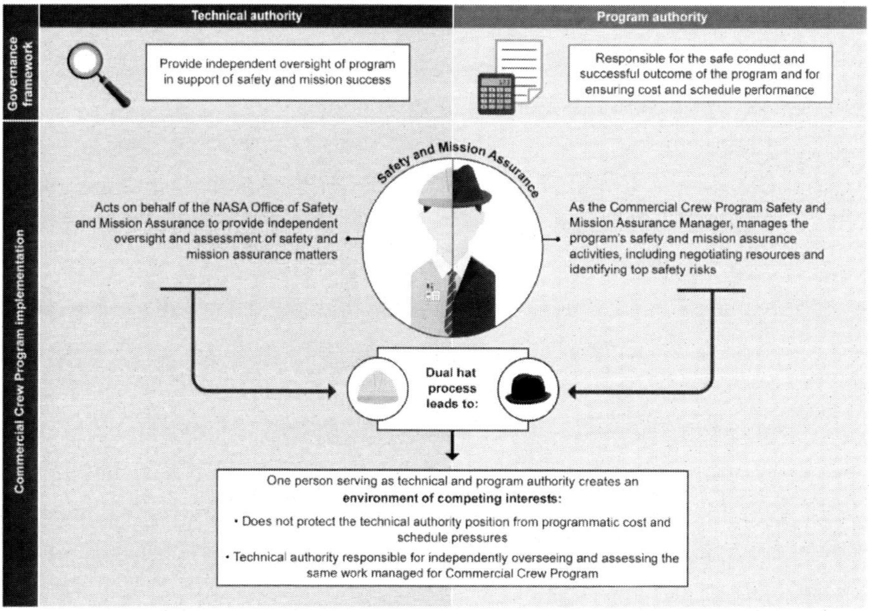

Source line: GAO analysis of National Aeronautics and Space Administration (NASA) data. | GAO-18-476.

Figure 5. Conflicting Roles and Responsibilities of Commercial Crew Program's Safety and Mission Assurance Technical Authority.

As a result, this structure relies on the same individual to completely separate two roles—one to manage the Commercial Crew Program's safety issues within programmatic cost and schedule constraints, and the other to assess the same issues in an independent oversight role. While the Commercial Crew Program may have an additional level of separation between the safety technical authority and the program's involvement in the design of commercial systems due to its shared assurance model with the commercial providers, the Commercial Crew Program still maintains a structure where one individual simultaneously serves in both technical authority and programmatic roles. Figure 5 describes some of the conflicting roles and responsibilities of this official in his two different positions.

During our review, officials cited several factors in support of a dual-hatted approach:

- The safety technical authority retains independence because his technical authority reporting path and performance reviews are not under the purview of the Commercial Crew Program chain of command.
- Due to the Commercial Crew Program's shared assurance model with commercial providers, the program is operating in a quality assurance role that provides an additional level of separation between the safety technical authority and the program's involvement in the design of commercial systems.
- For safety decisions involving cost and schedule where the individual who is dual hatted with both technical authority and programmatic responsibilities may feel conflicted, he stated that he would discuss these matters with his management to validate the logic behind his decision.
- There are cost and knowledge efficiencies gained from one individual serving in both programmatic and safety technical authority capacities.

NASA's Chief of Safety and Mission Assurance stated that he has great confidence in the individual currently serving in the dual hatted role for the Commercial Crew Program, but acknowledged there is inherent conflict even with the program's shared assurance model. In December 2017, he stated that, based on our previous work and current discussions, he intends to decouple the programmatic and technical authority responsibilities for the Commercial Crew Program but had not done so as of April 2018. Federal internal control standards state that an agency should design control activities to achieve objectives and respond to risks, which includes segregation of key duties and responsibilities to reduce the risk of error, misuse, or fraud.[22] By overlapping technical authority and programmatic responsibilities, NASA will continue to run the risk of creating an environment of competing interests for the Commercial Crew Program's safety technical authority.

---

[22] GAO-14-704G.

CONCLUSION

NASA's Commercial Crew Program is a multibillion dollar effort to facilitate the commercial development of a crew transportation system that can end the United States' reliance on Russia to maintain an uninterrupted presence on the ISS. Boeing and SpaceX continue to make progress developing a capability to fly to the ISS, but both have continued to experience delays. Program analysis indicates risks of further delays in each contractor's current schedule, but NASA has not provided that information to Congress in its routine briefings. Without this information, Congress does not know the full extent of potential delays to inform decision making. Additional delays could also disrupt U.S. access to the ISS. While NASA is working on potential solutions, there is no contingency plan in place to address this potential gap. Without a viable contingency plan, NASA puts at risk achievement of the U.S. goal and objective for the ISS.

NASA must balance safety with acceptable risk for human spaceflight. As part of the certification process for each contractor's spacecraft, NASA has developed one key safety metric, loss of crew. However, the complicated nature of this metric is further muddled by the inconsistent approaches being used across NASA about what inputs to be considered. As a result, there is no clear articulation of what level of risk NASA will accept with respect to this program. In addition, NASA does not have plans to capture lessons learned from how the Commercial Crew Program has used this metric to assess safety and is missing an opportunity to capture this knowledge for future human spaceflight programs.

Finally, a space program's management and oversight approach is an integral part of ensuring that human spaceflight is as safe and successful as possible. Independence of the program management and oversight functions is key to achieving the balance between safety and success. The Commercial Crew Program's approach, however, burdens the safety technical authority with both programmatic and independent technical authority responsibilities. As a result, NASA has limited assurance that independence can be maintained as part of its institutional process to ensure safety and success.

RECOMMENDATIONS FOR EXECUTIVE ACTION

We are making the following five recommendations to NASA:

The NASA Associate Administrator for Human Exploration and Operations should direct the Commercial Crew Program to include the results of its schedule risk analysis in its mandatory quarterly reports to Congress. (Recommendation 1)

The NASA Administrator should develop and maintain a contingency plan for ensuring a presence on the ISS until a Commercial Crew Program contractor is certified. (Recommendation 2)

The NASA Administrator should direct the Chief of Safety and Mission Assurance, the NASA Associate Administrator for Human Exploration and Operations, the Commercial Crew Program Manager, and the Commercial Crew Program Contracting Officer to collectively determine and document before the agency certification review how the agency will determine its risk tolerance level with respect to loss of crew. (Recommendation 3)

After completing the agency certification review, NASA's Chief Engineer and Chief of Safety and Mission Assurance, with support from the NASA Associate Administrator for Human Exploration and Operations and the Commercial Crew Program Manager, should document lessons learned related to loss of crew as a safety threshold for future crewed spaceflight missions, given the complexity of the metric. (Recommendation 4)

The NASA Chief of Safety and Mission Assurance should restructure the technical authority within the Commercial Crew Program to ensure that the technical authority for the Office of Safety and Mission Assurance is no longer dual hatted with programmatic and independent technical authority responsibilities. (Recommendation 5)

AGENCY COMMENTS AND OUR EVALUATION

We provided a draft of this chapter to NASA for review and comment. NASA provided written comments that are reprinted in appendix II.

In its response, NASA concurred with three of our recommendations, did not concur with one, and partially concurred with another.

- NASA concurred with our recommendation to develop and maintain a contingency plan to ensure a U.S. presence on the ISS and expects to take action to close this recommendation by the end of December 2018.
- NASA concurred with our recommendation to document lessons learned related to the loss of crew requirement and expects to take action to close this recommendation by the end of May 2019.
- NASA concurred with our recommendation to restructure the safety technical authority so that it is no longer dual hatted with programmatic and independent technical authority responsibilities. NASA expects to take action to close this recommendation by the end of August 2018.

NASA did not concur with our recommendation that the Commercial Crew Program should include the results of its schedule risk analysis in its quarterly reports to Congress. NASA stated that it uses the contractors' schedules as a baseline to provide qualitative statements in the NASA summary that accompanies each contractor's quarterly reports to Congress. NASA believes that this approach is appropriate and is in accordance with the explanatory statement accompanying the Consolidated and Further Continuing Appropriations Act, 2015. NASA also stated that it will be working to ensure that the contractors' schedules and the program's internal assessments sync up as the program gets closer to launch. As a result, NASA explained that there will not be a requirement for a detailed NASA assessment, because the contractors' schedule will either match NASA's analysis or NASA will discuss its position as it has done in previous reports to Congress.

We continue to believe the recommendation is valid because the program's schedule risk analysis would provide Congress with valuable insight into potential delays, which are likely. Both contractors have repeatedly stated that their schedules are aggressive and that the dates are ambitious. As a result, we found that the contractors frequently delay dates for key events. For example, Boeing has delayed its certification milestone by 17 months and SpaceX by 22 months since the original schedules were established. The program's recent schedule risk analysis indicates that more delays to certification are likely, but that information is not presented to Congress in NASA's quarterly reports. Without this information, Congress does not know the full extent of potential delays to inform decision making.

NASA partially concurred with our recommendation that the Chief of Safety and Mission Assurance, the NASA Associate Administrator for Human Exploration and Operations, the Commercial Crew Program Manager, and the Commercial Crew Program Contracting Officer should collectively determine and document how the agency will determine its risk tolerance level with respect to loss of crew before the agency certification review. In its response, NASA stated that it documented the agency's risk tolerance level with respect to loss of crew for the program in its May 2011 safety memo. Further, NASA stated that it documented the requirement to limit risks to the loss of crew in a certification requirements document. NASA stated that ultimately the Commercial Crew Program is accountable for ensuring that the contractors' systems meet the loss of crew value in this certification requirements document, which is a loss of crew value of 1 in 270. If a contractor's system cannot meet that loss of crew value, or any other requirement, the program will request a waiver as part of the human rating certification process to ensure transparency.

NASA acknowledged in its response that the existence of multiple documents defining residual risk requirements and an agency threshold for loss of crew can be confusing. NASA's response, however, does not address our finding that it does not have a consistent approach for how to incorporate key inputs, including which debris model should be used or whether to include operational mitigations. NASA stated that it had taken action to address this recommendation; however, NASA did not outline any steps it took to resolve the concern that the risk tolerance for the loss of crew requirement depends on which entity is presenting the results of its analysis. We continue to believe that, before the agency certification review, the key parties must collectively determine how the agency will determine its risk tolerance with respect to loss of crew. We believe this approach will reduce confusion and increase transparency.

Cristina T. Chaplain
Director, Contracting and National Security Acquisitions

APPENDIX I: OBJECTIVES, SCOPE, AND METHODOLOGY

The objectives of our review were to assess (1) the extent to which the contractors have made progress towards meeting the National Aeronautics and Space Administration's (NASA) certification requirements and NASA's plans to help ensure continued access to the International Space Station (ISS); and (2) how NASA's certification process addresses safety of the contractors' crew transportation systems.

To assess the contractors' progress towards certification, we obtained and reviewed program and contractor documents, including monthly and quarterly updates from April 2017 through May 2018. We interviewed program and contractor officials to discuss the contractors' recent progress, including upcoming events and any expected delays, and to understand technical risks, potential consequences, and planned mitigation activities. To identify total delays to date, we compared original contract schedules to Boeing and SpaceX's calendar year 2018 first quarter proposed schedules, which are the most recent. Based on our review of program and contractor documents, we defined the contractors' key events as: the uncrewed and crewed flight tests, the design certification reviews for each of those flights, and the certification milestone. We selected the two flight tests for each contractor as key events because they are intended to test key system capabilities, including the ability to launch, dock with the ISS, and return safely to Earth. We selected the design certification reviews because they verify the contractors' crew transportation systems' capability to safely approach, dock, mate, and depart from the ISS, among other requirements. We selected the certification milestone because it determines whether the crew transportation system meets the Commercial Crew Program's requirements.

To determine the extent to which contractors have delayed these key events over time, we analyzed the contractors' schedule data from the 13 quarterly progress reports to date, from first quarter 2015 through first quarter calendar year 2018. We also obtained the results of the program's April 2018 schedule risk analysis. We presented the schedule analysis range from the end of the month of the earliest possible completion date to the end of the month of the latest possible completion date. We reviewed the program's Congressional requirements to report on cost, schedule, and technical status. Finally, to assess the potential effects of any certification delays on NASA's access to ISS, we reviewed NASA's contracts with Boeing and the Russian Federal Space Agency for transportation on the Soyuz vehicle. We interviewed officials from the ISS program and NASA's Human Exploration and Operations Mission Directorate to determine if the agency had developed contingency plans to mitigate the effects of any certification delays on its access to the ISS.

To assess how NASA's certification process addresses safety of the contractors' crew transportation systems, we reviewed agency safety policies, program plans, and contract documents to identify what safety assessments were required of the program, which safety factors would be considered in certification reviews, and when certification approval would be granted. We also interviewed program officials and the contractors about their safety policies and procedures as well as about the certification process. We identified the loss of crew requirements for the program and the contractors, and interviewed program and agency officials to determine how loss of crew would be assessed and considered throughout the certification process. To gain a broader understanding of the relative importance of loss of crew, we reviewed NASA safety policies, prior GAO reports, and annual reports from the Aerospace Safety Advisory Panel. To determine how the Orbital Debris Engineering Model (ORDEM) was updated and to assess differences between the former and updated models, we reviewed NASA documentation about the ORDEM 2000 and ORDEM 3.0 models and obtained information about the models from NASA's Orbital Debris Program Office.

To assess the program's progress in closing requirements, including those requirements specifically related to safety, we reviewed program data for each contractor on contract requirements and closure status. We limited this analysis to requirements included in the CCT-REQ-1130 *ISS Crew Transportation and Services Requirements Document* because these requirements are verified by the Commercial Crew Program, whereas requirements contained within SSP 50808 *ISS to Commercial Orbital Transportation Services Interface Requirements Document* are managed by the ISS Transportation and Integration Office. We classified requirements as "safety-specific" when the program's Office of Safety and Mission Assurance was listed as a verification closure notice signatory (i.e., approver). We then analyzed the program's data to determine how many verification closure notices had been approved for all requirements and for the subset of safety-specific requirements. We reviewed program and agency documentation, such as organizational charts, program plans, and safety policies as well as interviewed program and agency officials, to determine the role of the safety and mission assurance technical authority in the program. We also reviewed the 2003 Columbia Accident Investigation Board's Report's findings and recommendations related to culture and organizational management of human spaceflight programs. We reviewed annual briefings and reports and met with representatives from two organizations that provide NASA with independent assessments of the program, the program's standing review board and the Aerospace Safety Advisory Panel, to gain their perspectives on the contractor's progress and how NASA addresses safety in its certification process. We also met with representatives from the National Transportation Safety Board and three experts with background on safety in human spaceflight in order to increase our contextual understanding of the role of safety in human spaceflight missions.

We conducted this performance audit from April 2017 to July 2018 in accordance with generally accepted government auditing standards. Those standards require that we plan and perform the audit to obtain sufficient, appropriate evidence to provide a reasonable basis for our findings and conclusions based on our audit objectives. We believe that the evidence obtained provides a reasonable basis for our findings and conclusions based on our audit objectives.

APPENDIX II: COMMENTS FROM THE NATIONAL AERONAUTICS AND SPACE ADMINISTRATION

National Aeronautics and Space Administration
Headquarters
Washington, DC 20546-0001

JUN 20 2018

Reply to Attn of: Human Exploration and Operations Mission Directorate

Ms. Cristina T. Chaplain
Director
Contracting and National Security Acquisitions
United States Government Accountability Office
Washington, DC 20548

Dear Ms. Chaplain:

The National Aeronautics and Space Administration (NASA) appreciates the opportunity to review and comment on the Government Accountability Office (GAO) draft report entitled, "NASA Commercial Crew Program: Plan Needed to Ensure Uninterrupted Access to the International Space Station" (GAO-18-476) dated May 15, 2018.

In the draft report, GAO makes five recommendations intended to ensure continued access to the International Space Station (ISS). Specifically, GAO recommends the following:

Recommendation 1: The NASA Associate Administrator for Human Exploration and Operations should direct the Commercial Crew Program to include the results of its schedule risk analysis in its mandatory quarterly reports to Congress.

Management's Response: Non-Concur. NASA's process is to use the partners' schedules as the baseline and to provide a qualitative statement regarding milestone schedules in the NASA summary that accompanies each commercial partners' Congressional report. NASA believes this current reporting approach is appropriate and in accordance with the Explanatory Statement accompanying the FY 2015 Consolidated and Further Continuing Appropriations Act (P.L. 113-235). NASA is protecting for future schedule adjustments (see Management Response to recommendation #2). As we are now getting closer to launch, we expect and will be working to ensure that the partners' schedules and NASA's internal assessments are in agreement. There will not be a requirement for detailed NASA risk assessment. The partners schedule risk assessment will match NASA's analysis or NASA will discuss our position as we have in past reports.

Recommendation 2: The NASA Administrator should develop and maintain a contingency plan for ensuring a presence on the ISS until a Commercial Crew Program contractor is certified.

Management's Response: Concur. NASA is actively developing, assessing, and implementing options to protect for presence of crew on the ISS to support the United States On-orbit Segment (USOS). The ISS Program continues to work with the Commercial Crew Program (CCP) to monitor launch schedules and is developing options to provide schedule margin to the CCP by protecting USOS presence if current launch dates are delayed beyond the return of U.S. crews on Soyuz.

Estimated Completion Date: December 31, 2018.

Recommendation 3: The NASA Administrator should direct the Chief of Safety and Mission Assurance, the NASA Associate Administrator for Human Exploration and Operations, the Commercial Crew Program Manager, and the Commercial Crew Program Contracting Officer to collectively determine and document before the agency certification review how the agency will determine its risk tolerance level with respect to loss of crew.

Management's Response: Partially concur. NASA has documented the Agency's risk tolerance level with respect to loss of crew for CCP missions in the Decision Memorandum for the Administrator, "Agency's Safety Goals and Thresholds for the Crew Transportation Missions to the International Space Station (ISS)," dated May 17, 2011. In addition, NASA has documented the requirements, standards, and certification package contents that will be used to certify a commercial crew transportation system to carry NASA or NASA-sponsored crewmembers on LEO missions, including the requirement to limit risks to the loss of crew, in HEOMD CCTS Certification Requirements for NASA LEO Missions (HEOMD-CSD-10001, Rev. B).

NASA recognizes that the existence of multiple documents defining residual risk requirements and an agency risk threshold for loss of crew can be confusing. Ultimately, the Commercial Crew Program is accountable for ensuring that the commercial systems meet the operational and design certification technical requirements defined in HEOMD-CSD-10001, including limiting the risk to loss of crew to 1 in 270. The CCP will document the commercial system's ability to meet that and other HEOMD-CSD-10001 requirements through design and operational mitigations in its Human Rating Certification Package. If the commercial system cannot meet that requirement (or any other requirements), the program will request a waiver from NASA Headquarters as part of the Human Rating Certification process to ensure transparency and to continue emphasis of safety performance under its contracts.

Estimated Completion Date: Action Completed.

Recommendation 4: After completing the Agency certification review, NASA's Chief Engineer and Chief of Safety and Mission Assurance, with the support from the NASA Associate Administrator for Human Exploration and Operations and the Commercial Crew Program Manager, should document the lessons learned related to loss of crew as a safety threshold for future crewed spaceflight missions, given the complexity of the metric.

Management's Response: Concur. After the Certification Reviews, the Office of Safety and Mission Assurance, along with the Office of Chief Engineer, the Human Exploration and Operations Mission Directorate, and CCP, will document the lessons learned related to the loss of crew requirement and thresholds. These lessons learned will be applied as appropriate to the requirements for future crewed spaceflight missions.

Estimated Completion Date: May 31, 2019.

Recommendation 5: The NASA Chief of Safety and Mission Assurance should restructure the technical authority within the Commercial Crew Program to ensure that the technical authority for the Office of Safety and Mission Assurance is no longer dual-hatted with the programmatic and independent technical authority responsibilities.

Management's Response: Concur. The NASA Chief of Safety and Mission Assurance has reviewed the implementation of the Safety and Mission Assurance (SMA) technical authority for CCP. The SMA programmatic responsibilities and the independent technical authority responsibilities are separated into two unique and distinct positions in the organization. The CCP SMA Plan, CCT-PLN-3100, is in the process of being updated to reflect and clarify this separation of responsibilities.

Estimated Completion Date: August 31, 2018.

Sincerely,

William H. Gerstenmaier
Associate Administrator
for Human Explorations and Operations

Ralph R. Roe, Jr.
Chief Engineer

Terrence W. Wilcutt
Chief, Safety and Mission Assurance

APPENDIX III: STAFF ACKNOWLEDGMENTS

In addition to Cristina T. Chaplain, Molly Traci, Assistant Director; Kazue Chinen; Lorraine Ettaro; Lisa Fisher; Laura Greifner; Kurt Gurka; Miranda Riemer; Juli Steinhouse; Roxanna T. Sun; Hai Tran; Kristin Van Wychen; and Alyssa Weir made significant contributions to this chapter.

APPENDIX IV: ACCESSIBLE DATA

Agency Comment Letter

Text of Appendix II: Comments from the National Aeronautics and Space Administration

Dear Ms. Chaplain:

The National Aeronautics and Space Administration (NASA) appreciates the opportunity to review and comment on the Government Accountability Office (GAO) draft report entitled, "NASA Commercial Crew Program: Plan Needed to Ensure Uninterrupted Access to the International Space Station" (GAO-18-476) dated May 15, 2018.

In the draft report, GAO makes five recommendations intended to ensure continued access to the International Space Station (ISS). Specifically, GAO recommends the following:

Recommendation 1

The NASA Associate Administrator for Human Exploration and Operations should direct the Commercial Crew Program to include the results of its schedule risk analysis in its mandatory quarterly reports to Congress.

Management's Response

Non-Concur. NASA's process is to use the partners' schedules as the baseline and to provide a qualitative statement regarding milestone schedules in the NASA summary that accompanies each commercial partners' Congressional report. NASA believes this current reporting approach is appropriate and in accordance with the Explanatory Statement accompanying the FY 2015 Consolidated and Further Continuing Appropriations Act (P.L. 113- 235). NASA is protecting for future schedule adjustments (see Management Response to recommendation #2). As we are now getting closer to launch, we expect and will be working to ensure that the partners' schedules and NASA's internal assessments are in agreement. There will not be a requirement for detailed NASA risk assessment. The partners schedule risk assessment will match NASA's analysis or NASA will discuss our position as we have in past reports.

Recommendation 2

The NASA Administrator should develop and maintain a contingency plan for ensuring a presence on the ISS until a Commercial Crew Program contractor is certified.

Management's Response

Concur. NASA is actively developing, assessing, and implementing options to protect for presence of crew on the ISS to support the United States On-orbit Segment (USOS). The ISS Program continues to work with the Commercial Crew Program (CCP) to monitor launch schedules and is developing options to provide schedule margin to the CCP by protecting USOS presence if current launch dates are delayed beyond the return of U.S. crews on Soyuz.

Estimated Completion Date

December 31, 2018.

Recommendation 3

The NASA Administrator should direct the Chief of Safety and Mission Assurance, the NASA Associate Administrator for Human Exploration and Operations, the Commercial Crew Program Manager, and the Commercial Crew Program Contracting Officer to collectively determine and document before the agency certification review how the agency will determine its risk tolerance level with respect to loss of crew.

Management's Response

Partially concur. NASA has documented the Agency's risk tolerance level with respect to loss of crew for CCP missions in the Decision Memorandum for the Administrator, "Agency's Safety Goals and Thresholds for the Crew Transportation Missions to the International Space Station (ISS)," dated May 17, 2011. In addition, NASA has documented the requirements, standards, and certification package contents that will be used to certify a commercial crew transportation system to carry NASA or NASA- sponsored crewmembers on LEO missions, including the requirement to limit risks to the loss of crew, in HEOMD CCTS Certification Requirements for NASA LEO Missions (HEOMD-CSD-10001, Rev. B).

NASA recognizes that the existence of multiple documents defining residual risk requirements and an agency risk threshold for loss of crew can be confusing.

Ultimately, the Commercial Crew Program is accountable for ensuring that the commercial systems meet the operational and design certification technical requirements defined in HEOMD-CSD-10001, including limiting the risk to loss of crew to 1 in 270. The CCP will document the commercial system's ability to meet that and other HEOMD-CSD-10001 requirements through design and operational mitigations in its Human Rating Certification Package. If the commercial system cannot meet that requirement (or any other requirements), the program will request a waiver from NASA Headquarters as part of the Human Rating Certification process to ensure transparency and to continue emphasis of safety performance under its contracts.

Estimated Completion Date

Action Completed.

Recommendation 4

After completing the Agency certification review, NASA's Chief Engineer and Chief of Safety and Mission Assurance, with the support from the NASA Associate Administrator for Human Exploration and Operations and the Commercial Crew Program Manager, should document the lessons learned related to loss of crew as a safety threshold for future crewed spaceflight missions, given the complexity of the metric.

Management's Response

Concur. After the Certification Reviews, the Office of Safety and Mission Assurance, along with the Office of Chief Engineer, the Human Exploration and Operations Mission Directorate, and CCP, will document the lessons learned related to the loss of crew requirement and thresholds. These lessons learned will be applied as appropriate to the requirements for future crewed spaceflight missions.

Estimated Completion Date

May 31, 2019.

Recommendation 5

The NASA Chief of Safety and Mission Assurance should restructure the technical authority within the Commercial Crew Program to ensure that the technical authority for the Office of Safety and Mission Assurance is no longer dual-hatted with the programmatic and independent technical authority responsibilities.

Management's Response

Concur. The NASA Chief of Safety and Mission Assurance has reviewed the implementation of the Safety and Mission Assurance (SMA) technical authority for CCP. The SMA programmatic responsibilities and the independent technical authority responsibilities are separated into two unique and distinct positions in the organization. The CCP SMA Plan, CCT-PLN-3100, is in the process of being updated to reflect and clarify this separation of responsibilities.

Estimated Completion Date

August 31, 2018.

Sincerely,

William H. Gerstenmaier
Associate Administrator for Human Explorations and Operations

Ralph R. Roe, Jr.
Chief Engineer

Terrence W. Wilcutt
Chief, Safety and Mission Assurance

In: Commercial Crew Program ...
Editor: Graeme N. Lutz
ISBN: 978-1-53614-980-7

Chapter 2

NASA COMMERCIAL CREW PROGRAM: CONTINUED DELAYS POSE RISKS FOR UNINTERRUPTED ACCESS TO THE INTERNATIONAL SPACE STATION*

Cristina T. Chaplain

WHY GAO DID THIS STUDY

Since the Space Shuttle was retired in 2011, the United States has been relying on Russia to carry astronauts to and from the space station. NASA's Commercial Crew Program is facilitating private development of a domestic system to meet that need safely, reliably, and cost-effectively before the seats it has contracted for on a Russian spacecraft run out in 2019.

---

* This is an edited, reformatted and augmented Accessible Version of United States Government Accountability Office Testimony Before the Subcommittee on Space, Committee on Science, Space, and Technology, House of Representatives, Publication No. GAO-18-317T, dated January 17, 2018.

In 2014, NASA awarded two firm-fixed price contracts to Boeing and SpaceX worth a combined total of up to $6.8 billion to develop crew transportation systems and conduct initial missions to the space station. In February 2017, GAO found that both contractors had made progress, but their schedules were under mounting pressure.

This statement provides preliminary observations on the extent to which the contractors and the program are making progress toward meeting NASA's standards for human spaceflight, a process called certification.

This statement is based on ongoing work and information contained in GAO's February 2017 report on this program (GAO-17-137). To do this work, GAO analyzed contracts, schedules, and other documentation.

What GAO Recommends

GAO is not making any new recommendations. In February 2017, GAO recommended that NASA develop a contingency plan to maintain access to the ISS beyond 2018, when its contract with Russia for seats on the Soyuz was scheduled to end. NASA agreed with this recommendation and purchased Soyuz seats through 2019.

What GAO Found

Both Boeing and Space Exploration Technologies (SpaceX) are making progress toward their goal of being able to transport American astronauts to and from the International Space Station (ISS). However, both continue to experience schedule delays. Such delays could jeopardize the ability of the National Aeronautics and Space Administration's (NASA) Commercial Crew Program to certify either company's option—that is, to ensure that either option meets NASA standards for human spaceflight—before the seats the agency has contracted for on Russia's Soyuz spacecraft run out in 2019. (See figure.)

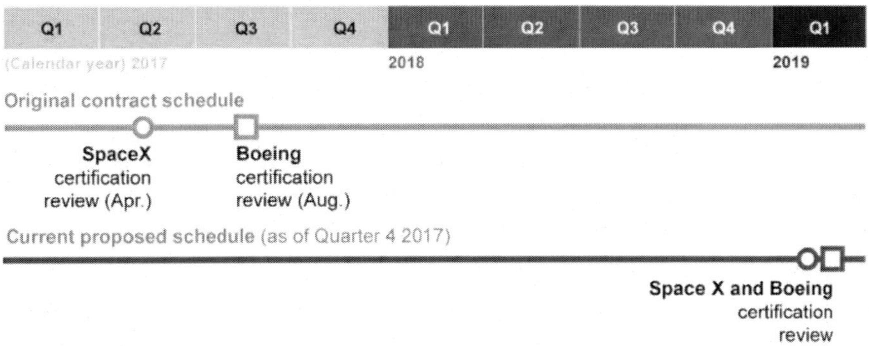

Source: GAO analysis of National Aeronautics and Space Administration contracts and documents. | GAO-18-317T.

Commercial Crew Program: SpaceX and Boeing's Certification Delays

GAO's ongoing work has identified three key risks, which are consistent with challenges reported in February 2017 that could further delay certification of each contractor's crew transportation system:

- *Aggressive schedules*—NASA, Boeing, SpaceX, and independent review bodies have all noted that the contractors' schedule plans are aggressive. The anticipated schedule risks have since materialized.
- *Programmatic and safety risks*—SpaceX and Boeing are addressing technical risks, which is not uncommon for NASA projects as they often push the state of the art in space technology. In addition, the contractors' systems must meet a standard for crew safety. Additional work remains to determine whether the contractors will meet this requirement.
- *Program office workload*—Program officials told GAO that one of their greatest upcoming challenges will be to complete two oversight activities—conducting phased safety reviews and verifying that contractors meet requirements—concurrently. The

program's ability to smooth its workload is limited, as the contractors generally control their development schedules. In February 2017, GAO found that proposed schedule changes could alleviate some overlap.

Delays and uncertain final certification dates raise questions about whether the United States will have uninterrupted access to the ISS after 2019, and may lessen NASA's return on investment with the contractors. GAO will continue to assess the contractors' and program's progress.

* * *

Chairman Babin, Ranking Member Bera, and Members of the Subcommittee:

I am pleased to be here today to discuss the status of the National Aeronautics and Space Administration's (NASA) Commercial Crew Program. As you know, following the retirement of the Space Shuttle in 2011, the United States was left with no domestic ability to provide crew access to the International Space Station (ISS). Since then, NASA has relied on purchasing seats from Russia on its Soyuz spacecraft to maintain a U.S. presence on the station. NASA's Commercial Crew Program is intended to end this dependency by facilitating the commercial development of a crew transportation system that can provide safe, reliable, and cost-effective transportation to and from low earth orbit, including the ISS. NASA's goal is to have one or more contractors that can provide crew transportation services to the ISS, which NASA expects will be operational until at least 2024.

NASA's acquisition strategy on the Commercial Crew Program is similar to the one it used on the Commercial Cargo program, but different than every other spacecraft it has built for humans, from Mercury to Gemini and Apollo to the Space Shuttle. For the Commercial Crew Program, each contractor designs, develops, builds, owns, and operates its spaceflight system and infrastructure. The contractors have access to NASA's expertise and resources throughout the development process, but

NASA engineers are not making design decisions, and NASA personnel are less involved in processing, testing, launching, and operating the crew transportation system. In the end, NASA will buy a crew transportation service—a ride for its astronauts to and from the ISS—much like it does for ISS cargo.

In the most recent phase of the Commercial Crew Program, NASA awarded firm-fixed-price contracts in 2014 to Boeing and Space Exploration Technologies Corporation (SpaceX), valued at up to $4.2 billion and $2.6 billion, respectively, for the development of crew transportation systems that meet NASA requirements and for flying initial missions to the ISS. According to the contracts, the companies were supposed to provide NASA all the evidence the agency needed to certify that their systems met its performance and safety requirements by 2017.

We have reviewed the Commercial Crew Program for the last two years as part of our annual assessment of all NASA's major projects.[1] We also issued a report on the program in February 2017 in response to a provision in the House Committee on Appropriations report accompanying H.R. 2578.[2] Prior to this, we issued a report in December 2011 that assessed NASA's acquisition approach for acquiring commercial crew transportation.[3] Over this time, we have reported that the program made progress as both contractors make progress finalizing their designs and building hardware, but we also found that schedule pressure has increased as contractors delay key events.[4]

My statement today provides our preliminary observations on the extent to which the contractors and the Commercial Crew Program are making progress in developing crew transportation systems that meet NASA's standards for human spaceflight, a process called certification.

---

[1] GAO, *NASA: Assessments of Major Projects*, GAO-17-303SP (Washington, D.C.: May 16, 2017); and *NASA: Assessments of Major Projects,* GAO-16-309SP (Washington, D.C.: Mar. 30, 2016).

[2] GAO, *NASA Commercial Crew Program: Schedule Pressure Increases as Contractors Delay Key Events*, GAO-17-137 (Washington, D.C.: Feb. 16, 2017).

[3] GAO, *National Aeronautics and Space Administration: Acquisition Approach for Commercial Crew Transportation Includes Good Practices, but Faces Significant Challenges*, GAO-12-282 (Washington, D.C.: Dec. 15, 2011).

[4] GAO-17-137.

This statement today is based upon our most recent report issued in February 2017 and some updated information since that report was published, which is based on ongoing work.[5] Our ongoing work is in response to a provision included in the house report accompanying H.R. 5393 for GAO to review the progress of NASA's human exploration programs.

For our ongoing work, to assess the extent to which the contractors are making progress toward certification, we obtained and reviewed program and contractor documents, including quarterly updates as well as monthly schedule summaries, from April 2017 through November 2017. We interviewed contractor officials to discuss the contractors' recent progress as well as their upcoming events and any expected delays. To identify total delays to date, we compared original contract schedules to Boeing's October 2017 working schedule and SpaceX's November 2017 working schedule, which identify their most recent proposed delays to some milestones. We also identified key risks facing the contractors and program by obtaining and reviewing monthly and quarterly reports, as well as the risks tracked in the program's risk management system, from April 2017 through November 2017. We interviewed program and contractor officials with knowledge of the technical risks to understand the risks and potential impacts and how they are planning to mitigate those risks.

The work upon which this statement is based is being conducted in accordance with generally accepted government auditing standards. Those standards require that we plan and perform the audit to obtain sufficient, appropriate evidence to provide a reasonable basis for our findings and conclusions based on our audit objectives. We believe the evidence obtained provides a reasonable basis for our findings and conclusions based on our audit objectives. More information about the scope and methodology for our February 2017 report can be found in that report.

---

[5] GAO-17-137.

We plan to issue a final report on the Commercial Crew Program in spring 2018. NASA provided us technical comments on information that is included in this statement, which we incorporated as appropriate.

BACKGROUND

NASA's Commercial Crew Program is a multi-phased effort that began in 2010. Across the five phases, NASA has engaged several companies using both agreements and contract vehicles to develop and demonstrate crew transportation capabilities. As the program has passed through these phases, NASA has generally narrowed down the number of participants. The early phases of the program were under Space Act agreements, which is NASA's other transaction authority.[6] These types of agreements are generally not subject to the Federal Acquisition Regulation (FAR) and allow the government and its contractors greater flexibility in many areas. Under these Space Act agreements, NASA relied on the commercial companies to propose specifics related to their crew transportation systems, including their design, the capabilities they would provide, and the level of private investment. In these phases, NASA provided technical support and determined if the contractors met certain technical milestones. In most cases, NASA also provided funding.

For the final two phases of the program, NASA awarded FAR-based contracts. By using FAR-based contracts, NASA gained the ability to levy specific requirements on the contractors and procure missions to the ISS, while continuing to provide technical expertise and funding to the contractors. Under these contracts, NASA will also evaluate whether contractors have met its requirements and certify their final systems for use.

In September 2014, NASA awarded firm-fixed-price contracts to Boeing and SpaceX, valued at up to $4.2 billion and $2.6 billion, respectively, for the Commercial Crew Transportation Capability phase.

[6] This authority allows an agency to enter into agreements "other than" standard government contracts or other traditional mechanisms.

Under a firm-fixed price contract, the contractor must perform a specified amount of work for the price negotiated by the contractor and government. This is in contrast to a cost-reimbursement contract, in which the government agrees to pay the contractor's reasonable costs regardless of whether work is completed. Thus, under the fixed-price contracts, the contractors must generally bear the risk of cost overruns or schedule delays.

During this phase, the contractors will complete development of crew transportation systems that meet NASA requirements, provide NASA with the evidence it needs to certify that those systems meet its requirements, and fly initial crewed missions to the ISS. Under the contracts, NASA and the companies originally planned to complete the certification review for each system by 2017. Figure 1 shows the spacecraft and launch vehicles for Boeing and SpaceX's crew transportation systems.

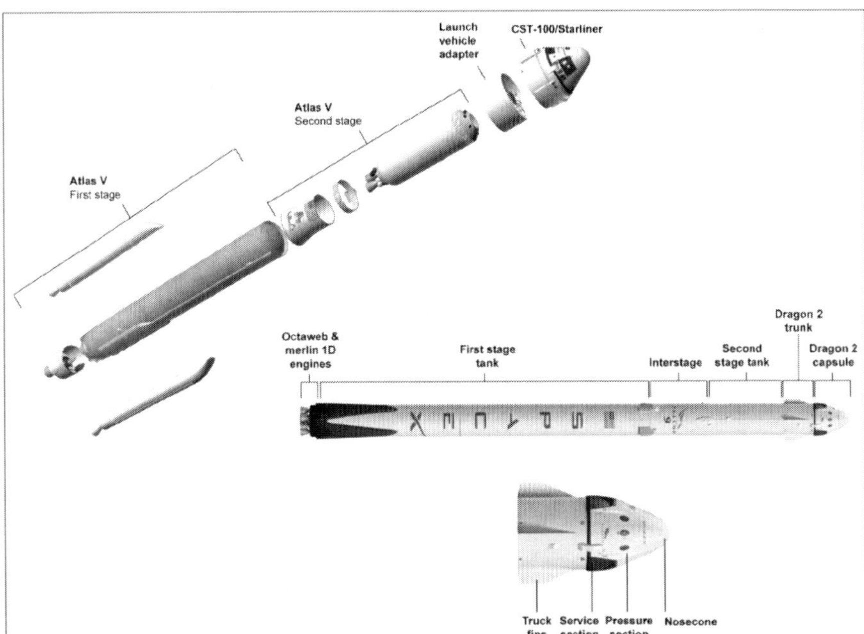

Source: ©2016 United Launch Alliance, LLC, (Top image); ©2016 SpaceX (bottom image). | GAO-18-317T.

Figure 1. Boeing and SpaceX Crew Transportation Systems.

The Commercial Crew Transportation Capability phase contracts include three types of services:

- *Contract Line Item 001* encompasses the firm-fixed-price design, development, test, and evaluation work needed to support NASA's final certification of the contractor's spacecraft, launch vehicle, and ground support systems.
- *Contract Line Item 002* covers any service missions that NASA orders to transport astronauts to and from the ISS. Under this indefinite-delivery, indefinite-quantity line item, NASA has ordered six missions from each contractor.[7] Each service mission is its own firm-fixed price task order. NASA must certify the contractors' systems before they can fly these missions.
- *Contract Line Item 003* is an indefinite-delivery, indefinite-quantity line item for any special studies, tests, and analyses that NASA may request. These tasks do not include any work necessary to accomplish the requirements under contract line item 001 and 002. As of July 2017, NASA had issued four orders under this contract line item to Boeing, worth approximately $1.8 million, including an approximately $180,000 study of the spacecraft's seat incline. NASA has issued one order under this contract line item to SpaceX, which did not affect the value of this line item. The maximum value of this contract line item is $150 million.

NASA divided the certification work under contract line item 001 into two acceptance events: the design certification review and the certification review. An acceptance event occurs when NASA approves a contractor's designs and acknowledges that the contractor's work is complete and meets the requirements of the contract. The design certification review verifies the contractor's crew transportation system's capability to safely

[7] An indefinite-delivery, indefinite-quantity contract may be used to acquire supplies or services during a specified contract period when the exact times and exact quantities of future deliveries are not known at the time of contract award.

approach, dock, mate, and depart from the ISS, among other requirements. After the contractor has successfully completed all of its flight tests, as well as various other activities, the certification review determines whether the crew transportation system meets the Commercial Crew Program's requirements. The contractors must complete both acceptance events to receive NASA certification.

NASA and the contractors also identified discrete performance-based events, called interim milestones, which occur as the contractors progress toward the two acceptance events. Each interim milestone has pre-determined entrance and exit criteria that establish the work that must be completed in order for the contractor to receive payment. The interim milestones serve several functions, allowing the government to finance work from development to completion, review the contractors' progress, and provide approval to proceed with key demonstrations and tests. The program also uses these milestones to inform its annual budget request. Since the contracts were awarded, the Commercial Crew Program and the contractors have agreed to split several of the interim milestones. The contractors have also added new milestones, in part to capture changes in their development plans.

NASA has also made changes to the contracts that have increased their value. While the contracts are fixed-price, their values can increase if NASA adds to the scope of the work or otherwise changes requirements. As of July 2017, NASA had increased the value of contract line item 001 for Boeing by approximately $48 million for hardware and software requirement changes, and contract line item 001 for SpaceX by approximately $91 million for a hardware requirement change and the addition of cargo during an ISS test flight.

In our February 2017 report,[8] we found the following:

- Both of the Commercial Crew Program's contractors have made progress developing their crew transportation systems, but both also have aggressive development schedules that are increasingly

[8] GAO-17-137.

under pressure. Both Boeing and SpaceX had determined that they would not be able to meet their original 2017 certification dates, and both expected certification to be delayed until 2018. We found that the schedule pressures were amplified by NASA's need to provide a viable crew transportation option to the ISS before its current contract with Russia's space agency runs out in 2019. If NASA needs to purchase additional seats from Russia, the contracting process typically takes 3 years. Without a viable contingency option for ensuring uninterrupted access to the ISS in the event of further Commercial Crew delays, we found that NASA was at risk of not being able to maximize the return on its multibillion dollar investment in the space station.

- The Commercial Crew Program was using mechanisms laid out in its contracts to gain a high level of visibility into the contractors' crew transportation systems, but maintaining the current level of visibility through certification could add schedule pressures. For example, due to NASA's acquisition strategy for this program, its personnel are less involved in the testing, launching, and operation of the crew transportation system. And while the program has developed productive working relationships with both contractors, the level of visibility that the program had required thus far had also taken more time than the program or contractors anticipated. Ultimately, the program has the responsibility for ensuring the safety of U.S. astronauts, and its contracts give it deference to determine the level of visibility required to do so. Moving forward though, we found that the program office could face difficult choices about how to maintain the level of visibility it feels it needs without adding to the program's schedule pressures.

In order to ensure that the United States had continued access to the ISS if the Commercial Crew Program's contractors experienced additional schedule delays, we recommended that the NASA Administrator develop a contingency plan for maintaining a presence on the ISS beyond 2018, including options to purchase additional Russian Soyuz seats, and report to Congress on the results. NASA concurred with this recommendation, and in February 2017, NASA executed a contract modification to procure an option for three crewmember seats from Boeing on the Russian Soyuz vehicle. Our analysis found that these seats represented a contingency plan for U.S. access to the ISS through 2019. In April 2017, NASA informed the Congress of this action.

Both Contractors Have Made Progress but Continue to Experience Schedule Delays

Contractors Continue to Advance Development of Their Crew Transportation Systems

Both Boeing and SpaceX have continued to make progress finalizing their designs and building hardware as they work toward final certification of their crew transportation systems, since we last reported in February 2017.[9] Each contractor's system includes a spacecraft and a launch vehicle with supporting ground systems. The contractors are also manufacturing test articles and flight spacecraft to support the uncrewed and crewed flight tests. The contractors plan to use the test articles to demonstrate system performance and the flight spacecraft to demonstrate their ability to meet contract requirements.

As table 1 shows, these test articles and flight spacecraft are currently in varying stages of completion—some are completed and in testing while others are still early in the manufacturing phase. Should any issues arise during integration and test or the flight tests planned for 2018, the contractors may have to complete rework on the spacecraft already under construction.

[9] GAO-17-137.

Table 1. Description of Boeing and SpaceX Hardware, Current Status, and Upcoming Events as of Fourth Quarter Calendar Year 2017

Spacecraft (name and type)	Purpose	Current status	Upcoming events
Boeing Flight spacecraft 1	Environmental testing 2018 Quarter (Q) 4 crewed flight test 2019 Q3 second postcertification mission	Crew module: constructed and integrated Service module: in construction	Boeing plans to conduct environ-mental testing starting in spring 2018 to test the spacecraft in conditions that simulate the space environment.
Boeing Flight spacecraft 2	2018 Q3 uncrewed flight test 2019 Q2 first post-certification mission	Crew module: in construction Service module: in construction	Boeing plans to join the crew and service modules together in the first quarter of 2018.
Boeing Test article 1	Validate effectiveness of spacecraft design and abort system	Crew module: constructed and integrated Service module: constructed and integrated	This test article has undergone testing throughout 2017.
Boeing Test article 2	Support ground tests 2018 Q2 pad abort test	Crew module: constructed and integrated Service module: in construction	This test article is completing testing before it will be reconfigured to support the pad abort test.
SpaceX Flight spacecraft 1	2018 Q3 uncrewed flight test	Crew module: constructed and integrated Support module:[a] in construction	SpaceX plans to join the crew and support modules together in the second quarter of 2018.
SpaceX Flight spacecraft 2	2018 Q4 crewed flight test	Crew module: in construction Support module: in construction	SpaceX plans to join the crew and support modules together in the third quarter of 2018.
SpaceX Flight spacecraft 3	2019 Q2 first post-certification mission	Crew module: in construction Support module: not yet started	SpaceX plans to join the crew and support modules together in the first quarter of 2019.

Table 1. (Continued)

Spacecraft (name and type)	Purpose	Current status	Upcoming events
SpaceX Test article	Support spacecraft propulsion testing	Testing is underway to validate performance of spacecraft engine propulsion system.	SpaceX plans to complete this testing by the third quarter of 2018.

Source: GAO analysis of National Aeronautics and Space Administration and contractor documents. | GAO-18-317T.

[a] For the purposes of this report, we refer to the SpaceX's Dragon as the crew module— it is composed of a pressure section and a service section. We refer to SpaceX's trunk as the support module. According to SpaceX, it serves as the launch vehicle adapter, it includes solar arrays for on-orbit power, and guidance fins for escape abort scenarios.

Schedule Delays Continue, and Risks Remain to Final Certification Dates

The contractors have notified NASA that final certification dates have slipped to the first quarter of calendar year 2019 and, through our ongoing work, we have identified three key risk areas that could further delay certification of each contractor's crew transportation system. These areas are (1) the contractors' aggressive schedules, (2) programmatic and safety risks, and (3) Commercial Crew Program's workload. These are consistent with the challenges we found facing the contractors and program in our February 2017 report.[10]

Aggressive Schedules

Since the award of the current Commercial Crew contracts in September 2014, the program, Boeing, and SpaceX have all identified the contractors' delivery schedules as aggressive. Program officials told us that, from the outset, they knew delays were likely due to the developmental nature of the program. Multiple independent review bodies—including the program's standing review board, the Aerospace

[10] GAO-17-137.

Safety Advisory Panel, and the NASA Advisory Council-Human Exploration and Operations committee—also noted the aggressiveness of the contractors' schedules as they move toward certification.

In February 2017, we found that both contractors had notified NASA that they would not be able to meet the 2017 final certification dates originally established in their contracts and expected final certification to be delayed until 2018. Based on our ongoing work, we found that the contractors have notified NASA that these dates have slipped further to the first quarter of calendar year 2019. Figure 2 shows the original Boeing and SpaceX contract schedule and the current proposed schedule for each contractor.

However, the extent to which these schedules represent an accurate estimate of each contractor's final certification date is unclear for the following two reasons:

1. Each contractor provides schedule updates to the Commercial Crew Program at quarterly status reviews, and the dates frequently change. The program has held 12 quarterly reviews since each contract was awarded. Boeing has reported a delay six times and SpaceX has reported a delay nine times that included at least one key event identified in the timeline above at these quarterly reviews.
2. The Commercial Crew Program is tracking risks that both contractors could experience additional schedule delays and, based on our ongoing work, we found that the program's own analysis indicates that certification is likely to slip into December 2019 for SpaceX and February 2020 for Boeing. Each month, the program updates its schedule risk analysis, based on the contractors' internal schedules as well as the program's perspectives and insight into specific technical risks. The Commercial Crew Program manager stated that differences between the contractors' proposed schedules and the program's schedule risk analysis include the following:

- o The contractors are aggressive and use their schedule dates to motivate their teams, while NASA adds additional schedule margin for testing.
- o Both contractors assume an efficiency factor in getting to the crewed flight test that NASA does not factor into its analysis.

The program manager explained further that the program meets with each contractor monthly to discuss schedules and everyone agrees to the relationships between events in the schedule even if they disagree on the length of time required to complete events. The program manager added, however, that she relies on her prior experience for a better sense of schedule timeframes as opposed to relying on the contractors' schedules.

While NASA has a fixed-price contract with both SpaceX and Boeing, there are consequences to the delays to date and the lack of certainty surrounding the final certification date. The United States has spent tens of billions of dollars to develop, assemble, and operate the ISS over the past two decades, and NASA relies on uninterrupted crew access to help maintain and operate the station itself and conduct the research required to enable human exploration in deep space and eventually Mars, among other science and research goals. To ensure uninterrupted access to the ISS through 2019, which includes launch and return of the astronauts, NASA purchased five seats on the Soyuz spacecraft through Boeing for an undisclosed value.[11] Boeing obtained these seats though a legal settlement with the Russian firm, RSC Energia, which manufactures the Soyuz. The NASA Office of Inspector General found in its annual report on NASA's top management and performance challenges that if the Commercial Crew Program experiences additional delays, NASA may need to buy additional seats from Russia to ensure a continued U.S. presence on the ISS.[12] Further, the ISS is planned to be operational through 2024. Unless there is a decision to extend the ISS's operational life, additional delays by Boeing

[11] In 2015, NASA paid approximately $82 million per seat through its contract with the Russian Federal Space Agency (Roscocosmos). See GAO-17-137.

[12] National Aeronautics and Space Administration, Office of Inspector General, *NASA's 2017 Top Management and Performance Challenges*, November 2017 (Washington, D.C.: November 2017).

and SpaceX may lessen NASA's return on investment with the contractors. We will continue to monitor this as part of our ongoing work.

	Boeing key events	Date	SpaceX key events
Current proposed schedule	Certification review	2019 Quarter (Q) 1	Certification review
Current proposed schedule	Crewed flight test	2018 Q4	Crewed flight test
Current proposed schedule	Uncrewed flight test	2018 Q3	Crewed design certification review Uncrewed flight test
Current proposed schedule	Crewed design certification review	2018 Q2	Uncrewed design certification review
Current proposed schedule		2018 Q1	
Current proposed schedule	Uncrewed design certification review	2017 Q4	
Original contract schedule	Certification review	2017 Q3	
Original contract schedule	Crewed flight test	2017 Q2	Certification review
Original contract schedule	Crewed design certification review Uncrewed flight test	2017 Q1	
Original contract schedule	Uncrewed design certification review	2016 Q4	Crewed flight test
Original contract schedule		2016 Q3	Crewed design certification review
Original contract schedule		2016 Q2	
Original contract schedule		2016 Q1	Uncrewed flight test
Original contract schedule		2015 Q4	Uncrewed design certification review

Source: GAO analysis of National Aeronautics and Space Administrations contracts and documents. | GAO-18-317T.

Figure 2. Boeing and SpaceX's Proposed Commercial Crew Schedule Delays as of Fourth Quarter Calendar Year 2017.

Programmatic and Safety Risks

In addition to challenges facing Boeing and SpaceX's aggressive schedules, both contractors face other risks that will need to be addressed to support their certification. This includes the contractors' ability to meet the agency's requirements related to the safety of their systems. These risks are not unusual; there are inherent technical, design, and integration risks in all NASA's major acquisitions, as these projects are highly complex and specialized and often push the state of the art in space technology. The Commercial Crew Program monitors risks through two lenses—programmatic risks potentially affect the program's cost and schedule or the performance of the crew transportation system, and safety risks could elevate the potential for the loss of crew.

SpaceX Risks

Similar to our findings in February 2017, our ongoing work indicates that the Commercial Crew Program's top programmatic and safety risks for SpaceX, are in part, related to ongoing launch vehicle design and development efforts.[15]

SpaceX must close several of the program's top risks related to its upgraded launch vehicle design, the Falcon 9 Block 5, before it can be certified for human spaceflight. Included in this Block 5 design is SpaceX's redesign of the composite overwrap pressure vessel. SpaceX officials stated the new design aims to eliminate risks identified in the older design, which was involved in an anomaly that caused a mishap in September 2016. Separately, SpaceX officials told us that the Block 5 design also includes design changes to address cracks in the turbine of its engine identified during development testing.

NASA program officials told us that they had informed SpaceX that the cracks were an unacceptable risk for human spaceflight. SpaceX officials told us that they have made design changes, captured in this Block 5 upgrade, that did not result in any cracking during initial life testing. However, this risk will not be closed until SpaceX successfully completes

[15] GAO-17-137.

qualification testing in accordance with NASA's standards without any cracks. SpaceX officials stated they expect this testing to be completed in first quarter calendar year 2018.

Finally, both the program and a NASA advisory group consider SpaceX's plan to fuel the launch vehicle after the astronauts are on board the spacecraft to be a potential safety risk. SpaceX's perspective is that this operation may be a lower risk to the crew. To better understand the propellant loading procedures, the program and SpaceX agreed to demonstrate the loading process five times from the launch site in the final crew configuration prior to the crewed flight test.

Boeing Risks

Our ongoing work indicates that Boeing is mitigating several risks in order to certify its crew transportation system, including challenges related to its abort system performance, parachutes, and its launch vehicle.

Boeing is addressing a risk that its abort system, which it needs for human spaceflight certification, may not meet the program's requirement to have sufficient control of the vehicle through an abort. In some abort scenarios, Boeing has found that the spacecraft may tumble and that could pose a threat to the crew's safety. To validate the effectiveness of its abort system, Boeing has conducted extensive wind tunnel testing and plans to complete a pad abort test in April 2018.

Boeing is also addressing a risk that during re-entry to the Earth's atmosphere, a portion of the spacecraft's forward heat shield may reconnect and damage the parachute system.[16] NASA's independent analysis indicates that this may occur if both parachutes that pull the forward heat shield away from the spacecraft deploy as expected. Boeing's analysis indicates the risk exists only if one of two parachutes does not deploy as expected. If the program determines this risk is unacceptable, Boeing would need to redesign the parachute system, which the program estimates could result in at least a 6-month delay. Finally, one of the program's top programmatic and safety concerns is that it may not have

[16] The forward heat shield protects the parachute system during re-entry.

enough information from Boeing's launch vehicle provider, United Launch Alliance, to assess if the launch vehicle prevents or controls cracking that could lead to catastrophic failures. The program and Boeing are in the process of negotiating next steps.

Program Safety Risk

The Commercial Crew Program has identified the ability of it and its contractors to meet a crew safety requirement as one of its top risks. NASA established the "loss of crew" metric as a way to measure the safety of a crew transportation system. The metric captures the probability of death or permanent disability to one or more crew members. Under each contract, the current loss of crew requirement is 1 in 270, meaning that the contractors' systems must carry no more than a 1 in 270 probability of incurring loss of crew. Near the end of the Space Shuttle program, the probability of loss of crew was approximately 1 in 90. As part of our ongoing work, we continue to work with NASA to understand how the loss of crew requirement was established for the Commercial Crew Program.

Program officials told us that Commercial Crew is the first NASA program that the agency will evaluate against a probabilistic loss of crew requirement. They said that if the contractors cannot meet the loss of crew requirement at 1 in 270, NASA could still certify their systems by employing operational mitigations. They said this would entail a potentially increased level of risk or uncertainty related to the level of risk for the crew.

Program officials told us their main focus is to work with the contractors to ensure that the spacecraft designs are robust from a safety perspective. The loss of crew metric and the associated models used to measure it are tools that help achieve that goal. For example, Boeing told us that in early 2016, it needed to identify ways to reduce the mass of its spacecraft. As Boeing found opportunities to reduce the spacecraft mass, the program stated that it had to consider how implementing those design changes would affect its loss of crew analysis in addition to compliance with other performance and safety requirements. According to the program, it is working with both contractors to address the factors that drive loss of crew risk through design changes or additional testing to gain more information on the performance and reliability of systems. As part of our ongoing work, we will continue to assess the extent to which the contractors are meeting this requirement and what tools the program and NASA will use to determine if the contractors meet the requirement.

Program Office Workload

In February 2017, we found that the Commercial Crew Program was using contractually defined mechanisms to gain a high level of visibility into the contractors' crew transportation systems, but also found that the Commercial Crew Program's workload was an emerging schedule risk.[17] At that time, program officials told us that one of their greatest upcoming challenges will be to keep pace with the contractors' schedules so that the program does not delay certification. Specifically, they told us they are concerned about an upcoming "bow wave" of work because the program must complete two oversight activities—phased safety reviews and verification closure notices—concurrently in order to support the contractors' design certification reviews, uncrewed and crewed flight test missions, and final certification.

The Commercial Crew Program is working to complete its three-phased safety review, which will ensure that the contractors have identified all safety-critical hazards and implemented associated controls, but it is behind schedule. Both the contractors and the program have contributed to these delays.

- In phase one, Boeing and SpaceX identified risks in their designs and developed reports on potential hazards, the controls they put in place to mitigate them, and explanations for how the controls will mitigate the hazards.

[17] GAO-17-137.

- In phase two, which is ongoing, the program reviews and approves the contractors' hazard reports, and develops strategies to verify and validate that the controls are effective.
- In phase three, the contractors plan to conduct the verification activities and incrementally close the reports.

The Commercial Crew Program's review and approval of the contractors' hazard reports have taken longer than planned. The program originally planned to complete phase two in early 2016, but through our ongoing work, we have found that as of October 2017, neither contractor had completed this phase. At that time, Boeing had completed 90 percent and SpaceX had completed 70 percent of the Phase 2 reports.

The Commercial Crew Program's verification closure notice process, which is used to verify that the contractors have met all requirements, is one of the other key oversight activities and potential workload challenges for the program. The program is completing that process concurrently with the phased safety reviews. The verification closure process is initiated by the contractor when it provides the program with data and evidence to substantiate that it has met each requirement, and is completed when the program has reviewed and approved the contractor's evidence to verify that each requirement has been met. The Commercial Crew Program must also approve a subset of verification closure notices before key tests or milestones can occur. For example, the ISS requirements and a portion of the Commercial Crew Program requirements must be met before Boeing and SpaceX's uncrewed flights to the ISS, which are currently planned for the third quarter of 2018. The program's ability to smooth its workload is limited because the contractors generally control their development schedules. In February 2017, we found, however, that proposed changes to the Boeing and SpaceX schedules could help alleviate some of the concurrency between the program's phased safety reviews and verification closure process.[18] We will continue to monitor the efforts as part of our ongoing work.

---

[18] GAO-17-137.

In conclusion, Boeing and SpaceX continue to make progress developing crew transportation systems to help the United States re-establish its domestic ability to provide crew access to the ISS. But, when the current phase of the Commercial Crew Program began, there was widespread acknowledgment that the contractors' development and certification schedules were aggressive and the anticipated schedule risks have now materialized. Further, programmatic and safety risks remain with schedules that frequently change making a final certification date uncertain. Delays and uncertain final certification dates raise questions about whether the United States will have uninterrupted access to the International Space Station beyond 2019, and may lessen NASA's return on investment with the contractors. We look forward to continuing to work with NASA and this subcommittee as we assess the contractors' and program's progress to final certification.

Chairman Babin, Ranking Member Bera, and Members of the Subcommittee, this completes my prepared statement. I would be pleased to respond to any questions that you may have at this time.

Staff Acknowledgments

GAO staff who made key contributions to this statement include Cristina T. Chaplain, Director; Molly Traci, Assistant Director; Susan Ditto; Lisa Fisher; Laura Greifner; Juli Steinhouse; Roxanna Sun; and Kristin Van Wychen.

In: Commercial Crew Program …
Editor: Graeme N. Lutz
ISBN: 978-1-53614-980-7

Chapter 3

COMMERCIAL SPACE INDUSTRY LAUNCHES A NEW PHASE*

Bill Canis

ABSTRACT

Rockets, satellites, and the services they provide, once the domain of governments, are increasingly launched and managed by privately owned companies. Although private aerospace firms have contracted with federal agencies since the onset of the Space Age six decades ago, U.S. government policy has sought to spur innovation and drive down costs by expanding the roles of satellite manufacturers and commercial launch providers.

Global spending on space activity reached an estimated $323 billion in 2015. Of this amount, nearly 40% was generated by commercial space products and services and 37% by commercial infrastructure and support industries. The U.S. government—including national security agencies and the National Aeronautics and Space Administration (NASA)—accounted for about 14% of global spending; government spending by other countries was responsible for the remaining 10%.

* This is an edited, reformatted and augmented version of a Congressional Research Service report, Publication No. R44708, dated December 12, 2016.

The satellite and launch vehicle supply chains are global, with a small number of manufacturers. In 2015, global satellite manufacturing revenues were $6 billion; launches booked $2.6 billion in revenue. Ground stations—the largest part of the commercial space infrastructure—generated more than $100 billion in revenue, largely from geolocation and navigation equipment.

The face of the U.S. space industry is changing with a government shift toward use of fixed price contracts for commercial services, new entrants with new launch products, and an increase in the use of smaller satellites:

- NASA's commercial cargo program and other federal contracts are supporting the growth of the commercial launch industry, with less expensive rockets, some of which are planned to be reusable. Many of the new space-related companies are attracting rising levels of venture capital.
- Aggressive pricing by U.S. entrants is cutting into the international launch market once dominated by foreign providers.
- A renewed interest in low-cost satellites, some of which are small enough to be held in one hand, is prompting a range of start-ups and providing new accessibility to space by educational institutions, small businesses, and individual researchers.

In order to spur innovation and growth, the commercial space industry has been purposely insulated from some types of federal regulation often applied to other industries. Nevertheless, three broad federal issues will affect the industry's future development. One is the structure of federal regulation and management; those responsibilities currently are dispersed among many agencies, and there is congressional interest in reorganizing commercial space functions at NASA and the Departments of Defense, Commerce, Transportation, and State. A second issue is the extent to which U.S. export controls are hampering U.S. satellite industry sales abroad. Export controls have recently been revamped to enable export of more commercial space products and services, but impediments may remain to reestablishing U.S. space product competitiveness. A third concern is that new Federal Communications Commission (FCC) regulations allowing wireless communication providers to share spectrum previously dedicated to satellite transmissions may result in interference. The commission has pledged to continue studying the issue.

INTRODUCTION

Rockets, satellites, and the services they provide, once the domain of governments, are increasingly launched and managed by privately owned companies. Until 1982, the U.S. government launched all civil and commercial payloads into orbit, and U.S. launch vehicle manufacturers produced vehicles only under contract to the National Aeronautics and Space Administration (NASA) or the Department of Defense (DOD). Most of the satellites they carried into orbit were owned by U.S. or foreign government agencies. Now, commercial payloads are generally launched by private providers, and the payloads themselves are increasingly likely to be owned by private entities: Of the 576 U.S.-owned satellites currently in orbit, 286 were launched for commercial reasons and another 12 on behalf of academic users.[1]

The growth of the commercial space sector is a result of a deliberate shift in federal policy. The Commercial Space Launch Act of 1984[2] states the following:

> [T]he United States should encourage private sector launches, reentries, and associated services and, only to the extent necessary, regulate those launches, reentries, and services to ensure compliance with international obligations of the United States and to protect the public health and safety, safety of property, and national security and foreign policy interests of the United States.[3]

That law was amended in 2004 to provide that "the regulatory standards governing human space flight ... evolve as the industry matures so that regulations neither stifle technology development nor expose crew or space flight participants to avoidable risks as the public comes to expect greater safety for crew and space flight participants from the industry."[4]

---

[1] Union of Concerned Scientists, *UCS Satellite Database*, http://www.ucsusa.org/nuclear-weapons/space-weapons/ satellite-database#, viewed October 25, 2016.

[2] P.L. 98-575.

[3] P.L. 98-575, §2.

[4] §2(a) of Commercial Space Launch Amendments Act of 2004 (CSLAA), P.L. 108-492.

National security agencies have emphasized the importance of the commercial space industry, in particular the space industrial base.[5] To foster its growth, the commercial space industry has purposely been insulated from some types of federal regulation often applied to other industries. For example, the Commercial Space Launch Act directs the Secretary of Transportation to issue regulations affecting the design and operation of launch vehicles only to protect the safety of crew, thereby giving launch vehicle manufacturers and operators wide leeway in developing new rockets. An initial eight-year "learning period" for this limited regulation has been twice extended, most recently until 2023.[6]

The Commercial Space Industry

Global spending on space activity reached an estimated $323 billion in 2015. Of this amount, nearly 40% was generated by commercial space products and services and 37% by commercial infrastructure and support industries. The U.S. government—including military and national security agencies and NASA—accounted for about 14% of global spending, and government spending by other countries the remaining 10%.[7]

The commercial space industry has distinct subsectors. Most commercial payloads are placed in orbits around Earth by launch vehicles that have the thrust to escape Earth's gravity. Typically, the firms that provide commercial launch services also design and assemble the rockets they launch. The payloads, such as satellites and manned space capsules, are manufactured by other firms. Service providers (such as television broadcasters) may design and build their own satellites, or purchase them from third-party manufacturers. The ground stations that control and

---

[5] Department of Defense and Office of the Director of National Intelligence, *National Security Space Strategy*, (Unclassified Summary), January 2011, p. 4, http://www.defense.gov/Portals/1/features/2011/0111_nsss/docs/NationalSecuritySpaceStrategyUnclassifiedSummary_Jan2011.pdf.

[6] The initial eight-year "learning period" was enacted in CSLAA and most recently extended by the Commercial Space Launch Competitiveness Act of 2015, P.L. 114-90.

[7] Space Foundation, *The Space Report*, 2016, p. 16.

communicate with the payloads, which form an integral part of the spacecraft operations, may be operated by the launch services, the payload owners, or entirely separate entities.

Satellites

The satellite supply chain is global, with a small number of manufacturers. Satellites are custom-made for their users, using specialized parts, which require extensive testing to ensure they will operate in a space environment where replacement and repair are generally not options. Because there are a limited number of rocket launches each year, supply chain deadlines are governed by a satellite launch window; if that window is missed because of production or regulatory delays, it may be many months or even years before another launch vehicle will be available to carry the satellite into orbit. Minimizing the mass and physical dimensions of a satellite is critically important, because heavier or bigger satellites are likely to require larger, more costly launch rockets. Appendix A shows a range of existing satellites, their weight, and corresponding familiar objects that approximate the same mass.

The United States is the largest manufacturer of spacecraft, followed by the European Union and Russia. The global commercial satellite manufacturing sector, with $6 billion in revenues in 2015, attempted 86 launches carrying 262 spacecraft. Of these, 126 were satellites weighing less than 22 pounds (10 kilograms). The growing use of these smaller satellites has led to a doubling in the number of satellites launched. However, smaller satellites comprise a tiny portion of the mass of vehicles sent into space: Of 392 tons of satellites launched in 2015, the total weight of satellites under 22 pounds was barely half a ton.[8]

Regardless of their size and type, satellites generally utilize similar types of components and instruments (Figure 1), including the following:

---

[8] Ibid., p. 17.

- *Solar panels* generate electricity that is stored in onboard *rechargeable batteries*.
- A *propulsion tank* adjusts altitude and orbit control and assists in placing the satellite into final orbit; several *thrusters* assist with these maneuvers and may be used later to move the satellite into a new orbit or, when the satellite's useful life is over, into a "graveyard" orbit farther away from Earth.
- *Thermal control* is the skin that prevents damage to the internal components in the extreme temperatures of space. Satellites pass through a wide range of temperatures, ranging from about -300°F to +200°F. Certain components, such as batteries, have an optimum operational range of about +50°F to +85°F.[9] The skin is composed of a multilayer thermal blanket made of lightweight reflective films. In addition, a coated or mirrored optical solar reflector may be used, which acts like a radiator to keep the components from overheating due to solar energy and radiation.
- An *electronics system* controls satellite functions, such as flight operations, the direction in which the satellite points, and mathematical analysis.
- A *communications system* includes a transmitter, a receiver, *antennae* that send data to and receive instructions from ground stations, and an amplifier that produces high-power radio frequency signals.
- Specialized components are determined by the core mission of the satellite. For example, a remote sensing satellite might include an image sensor, a telescope, and a digital camera.

Satellites are placed into different types of Earth orbits depending on their planned use: low Earth orbit, medium Earth orbit, geosynchronous Earth orbit, and high Earth orbit (Figure 2). The altitude of the orbit determines how frequently the spacecraft orbits the Earth. Some types of

---

[9] James Wertz, David Everett, and Jeffrey Puschell, "Spacecraft Subsystems V: Structural and Thermal," in *Space Mission Engineering: The New SMAD* (Space Technology Library, Hawthorne, CA, 2011), p. 686, Table 22-9.

space activities require closer proximity to Earth for supply and maintenance and to provide higher-resolution images of Earth. Others operate farther in space to more effectively deploy their services. The International Space Station (ISS), for example, is in low Earth orbit, taking 90 minutes for a full orbit, while a television or weather satellite in geosynchronous orbit will take a full day. A satellite in geosynchronous orbit matches the Earth's rotational speed, so unlike other types of satellites, it will remain in the same place above the Earth and can therefore provide the same locations with weather monitoring or telecommunications.

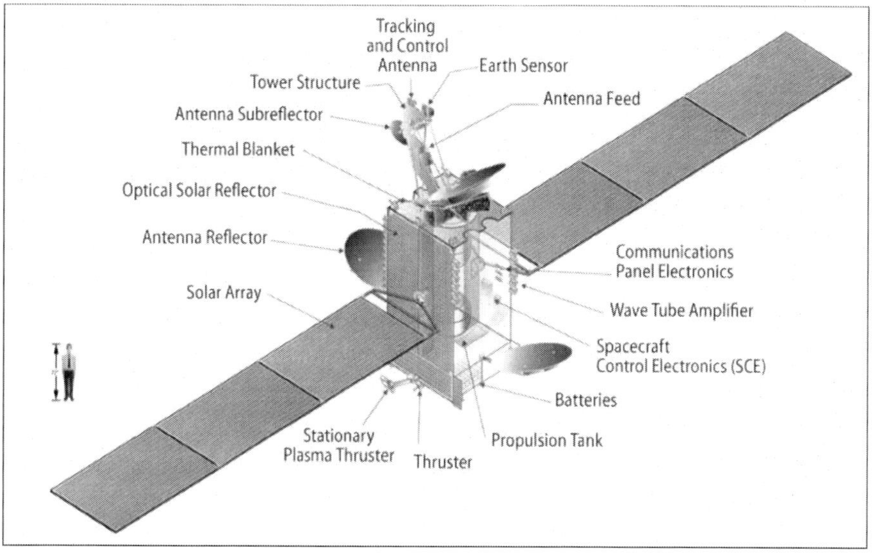

Source: Image courtesy of Space Systems Loral (SSL), http://www.sslmda.com.

Figure 1. Typical Satellite Components.

At the end of June 2016, 1,419 satellites were in operation, with 55% in low Earth orbit, 36% in geosynchronous orbit, 7% in middle Earth orbit, and the remainder in high Earth orbit. Of these, 576 are U.S. satellites, 140 Russian, 181 Chinese, and 522 from other countries. Of the U.S. satellites, 286 are commercial, 146 military, 132 government, and 12 civil.[10]

[10] Ibid.

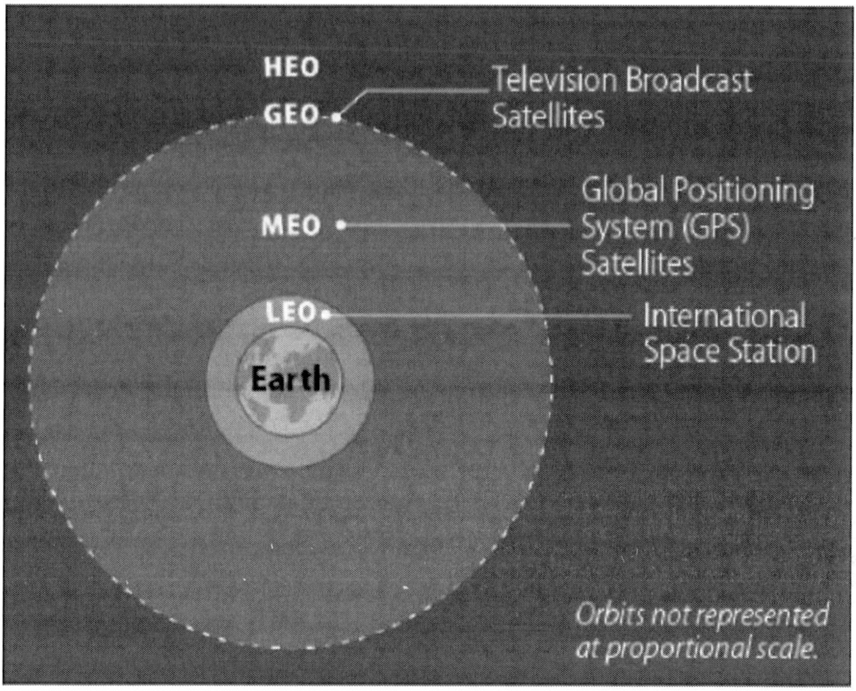

Source: CRS.

Figure 2. Earth Orbits.

Launch Services and Vehicles

The commercial launch industry booked $2.6 billion in revenues in 2015, when it attempted 86 rocket launches. Of these, 83 launches successfully placed payloads into orbit. Twenty-two launches carried commercial satellites and 64 carried government payloads; some carried both. It has been estimated that total costs of global launches for commercial, civil, and military purposes (including servicing the ISS) in 2015 were about $8 billion.[11] The U.S. Government Accountability Office (GAO) identified three factors spurring the growth of the U.S. launch industry:

[11] Ibid., p. 17.

- the NASA commercial cargo program and other federal contracts;
- aggressive pricing by SpaceX and some other private providers, which are "more price competitive compared with foreign launch providers"; and
- the emerging space tourism and small satellite industries.[12]

Launch Services

A traditional spaceport was a facility owned by the federal government, such as Kennedy Space Center and Cape Canaveral Air Force Station (both in Florida) and Vandenberg Air Force Base (in California). The federal government has used these facilities for its own launches and also promoted their use for commercial space access. Several private providers of launch services have chosen to build their own facilities. The Federal Aviation Administration (FAA) has licensed 10 spaceports in seven states: California, Florida, Texas, Oklahoma, Alaska, Virginia, and New Mexico.[13] California, Florida, and Texas each have two licensed spaceports. Additional launch sites have been proposed in other states.

Spaceports are configured for specific uses.[14] Some are planned only to launch large vertical rockets, while others hope to lure space tourism with facilities for winged launch vehicles.[15] Regardless of the type of launch vehicles using the sites, they have common infrastructure needs, including access for delivery of large launch vehicle components; room to assemble rocket parts into a launch vehicle; facilities for receiving and storing propellants and loading them aboard rockets; secure facilities for storing cargo, payloads, and scientific experiments; work space for crews,

---

[12] U.S. Government Accountability Office, *Commercial Space: Industry Developments and FAA Challenges*, GAO-16- 765T, June 22, 2016, pp. 11-12.

[13] Federal Aviation Administration, Active Launch Site Operator Licenses, September 17, 2013, https://www.faa.gov/ data_research/commercial_space_data/licenses/. FAA licensing authority is specified in 14 C.F.R. Part 420.

[14] Richard Rogers, "Commercial Spaceports: Building the Foundation of a Commercial Space Transportation Network," *TR News (Transportation Research Board)*, No. 300 (December 2015), p. 9.

[15] Some launch vehicles are expected to be shuttle-like vehicles, such as Virgin Galactic's SpaceShipTwo and Sierra Nevada's Dream Chaser.

engineers, and launch personnel; and meteorological equipment to monitor weather patterns prior to scheduled launches.[16]

The ISS[17] also increasingly offers certain types of launch services; part of it is designated as a national laboratory, a shared resource for NASA and private industry;[18] astronauts have used its robotic arm to launch small satellites into orbit, many of them only 4-inch cubes.[19] In 2015, 42 small satellites, primarily with commercial functions, were launched as cargo to the ISS for later deployment into individual orbits.[20] U.S. crew members and cargo were transported to the ISS in NASA's Space Shuttle from 2000 until 2011, when the shuttle was retired. Since then, the ISS has been resupplied by commercial launches through NASA's Commercial Orbital Transportation Services (COTS) program—the companies SpaceX and Orbital ATK are participants—as well as launches by the Japan Aerospace Exploration Agency (JAXA) and Russian Soyuz rockets.[21]

Launch Vehicles

Launch vehicles have the primary function of putting a spacecraft into an orbit or a suborbital trajectory. In the process of launching a satellite, most of the rocket stages fall away in sequence until the spacecraft reaches its planned orbit; the first stage of the rocket propels the rocket from the launch pad, and then the second-stage rocket boosts the payload to orbit.

---

[16] Other countries with spaceports are Australia, Brazil, China, EU, India, Iran, Israel, Japan, North Korea, Norway, Russia, South Korea, and Sweden. Space Foundation, *The Space Report*, 2016, p. 61.

[17] The ISS has been in operation since 2000 as a habitable satellite in LEO. It has a rotating multinational crew who conduct scientific research and demonstrate new technologies. NASA pays Russia $82 million for each roundtrip crew transport aboard Soyuz. NASA Office of Inspector General, *NASA's Commercial Crew Program: Update on Development and Certification Efforts*, IG-16-028, September 1, 2016.

[18] In 2005, Congress designated part of the ISS as a national laboratory in the NASA Authorization Act of 2005 (P.L. 109-155), §507.

[19] NASA, "Deploying Small Satellites From ISS," press release, November 10, 2015, https://www.nasa.gov/ mission_pages/station/research/benefits/cubesat.

[20] Federal Aviation Administration, *The Annual Compendium of Commercial Space Transportation*, January 2016, p. 36.

[21] In January 2016, NASA awarded a contract to Sierra Nevada Corporation to be third participant in its Commercial Resupply Services beginning in 2019. NASA, "NASA Awards International Space Station Cargo Transport Contracts," press release, January 14, 2016, https://www.nasa.gov/press-release/nasa-awards-international-space-station-cargo-transport-contracts.

Launch vehicles generally are used only once, although some commercial providers are developing vehicles that are intended to be reusable. Figure 3 is a cross section of a typical launch vehicle, the United Launch Alliance Atlas V. Its major components—from top to bottom—include the following:

- The *nose cone or fairing*, a structure made with a vented aluminum-honeycomb core and graphite epoxy covering, carries the payload. Manufacturers offer clients a choice of three payload fairings, depending on the size of the payload.[22] The fairing protects the payload from atmospheric pressure changes and aerodynamic heating during launch.
- The *second-stage rocket* consists of fuel and oxygen tanks, control systems, and a rocket engine that carries the payload to orbit. The Atlas V is propelled by a single RL 10 Centaur engine and stainless steel fuel tanks, providing 22,300 pounds-force (lbf)[23] of thrust, fueled by liquid hydrogen and liquid oxygen. The walls of the stainless steel tanks are insulated and so thin that they cannot support their own weight before they are pressurized, a design developed to maximize engine performance.[24] The Centaur second stage also includes flight and guidance computers that autonomously control all aspects of the flight.
- *Adapters* connect the first and second stages of the rocket and provide the structure for housing vehicle electronics.
- The *first stage* consists of additional fuel and oxygen tanks, control systems, and rocket engines, sometimes supplemented with strap-

---

[22] The height of an Atlas V rocket with a short fairing is 196 feet; with a medium fairing, 205 feet; and a long fairing, 215 feet. United Launch Alliance, http://www.ulalaunch.com/uploads/docs/Atlas500_Cutaway.pdf.

[23] The abbreviation *lbf* represents pound-force. As a measurement unit of force, it equals one pound multiplied by the standard acceleration due to gravity on Earth.

[24] The Centaur is the name of the second stage rocket used on Atlas rockets; variations of it have been in use since 1963. See http://www.nasa.gov/centers/glenn/about/history/centaur.html. The Aerojet Rocketdyne RL 10 engine has been used for more than 50 years as the second-stage engine on many commercial, civil, and military rockets. See http://www.rocket.com/rl10-engine.

on boosters. The Atlas V main booster is made of a special aluminum and, unlike the stage-two tanks, is structurally stable. The launch vehicle is fueled by rocket propellant (or highly purified kerosene) and liquid oxygen that provide 860,300 lbf of thrust. The RD 180 engine was developed in Russia and is produced by a U.S.-Russian joint venture.[25]

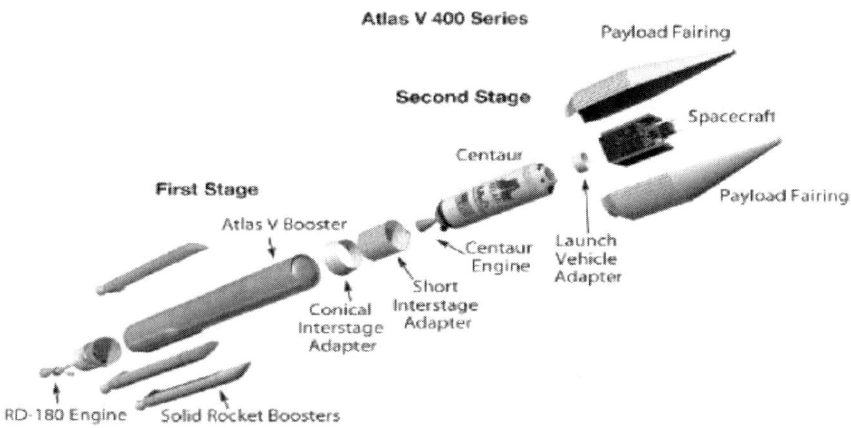

Source: Image courtesy of United Launch Alliance, http://www.ulalaunch.com/products_atlasv.aspx.

Figure 3. Cross Section of a Typical Launch Vehicle.

Ground Systems

Ground systems are the Earth-bound infrastructure that transmits directions to satellites and receives data they collect.

Ground systems include *antenna services* for transmission and reception of satellite radio frequency (RF) signals. There has been experimentation with optical communications (using lasers), but most satellite communication is through RF, generally using 30 megahertz

[25] The engines are produced in Russia by NPO Energomash and sold by the U.S.-Russian joint venture, known as RD AMROSS, a joint venture between Pratt & Whitney and NPO Energomash. See http://www.ulalaunch.com/faqs-rd180.aspx.

(MHz) to 30 gigahertz (GHz) bands.[26] Antenna services decode embedded data streams before passing them on for distribution and analysis.[27] Ground service options are customized for each satellite program and may include government-owned and commercial antenna assets to maximize a satellite's communications capability. A satellite that is in frequent contact with ground stations will need smaller amounts of onboard data storage than a spacecraft that has limited opportunities to transmit data to the ground.[28]

Ground facilities include *user terminals*, which can be handheld mobile terminals, rooftop dish antennae for satellite television, satellite radios in cars, or large corporate dish antennae.[29]

Ground systems also include *data accounting and distribution services* and *data processing* services. In addition to data storage, these services identify data missing from transmissions and take action to retrieve it from the satellite.[30] These services are highly automated "lights-out" systems that function around the clock with only occasional monitoring by employees. As the global space economy continues to expand, the need for international standards is growing to ensure interoperability among systems from different countries and service providers.[31]

With more than $100 billion in revenue, ground stations and related equipment comprise the largest part of commercial infrastructure. This subsector provides consumer products such as satellite phones and television, navigation chips in mobile phones, and many other products.

---

[26] Australian Space Academy, *Radio Frequencies for Space Communication*, http://www.spaceacademy.net.au/ spacelink/radiospace.htm.

[27] James Wertz, David Everett, and Jeffrey Puschell, "Spacecraft Subsystems V: Structural and Thermal," in *Space Mission Engineering: The New SMAD* (Space Technology Library, Hawthorne, CA, 2011), p. 879.

[28] Satellites with continual links to ground stations are built with no onboard storage. Ibid., pp. 880 and 897.

[29] Roger Cochetti, *Mobile Satellite Communications Handbook*, 2nd ed. (Hoboken, NJ: Wiley, 2015), pp. 31.

[30] James Wertz, David Everett, and Jeffrey Puschell, "Spacecraft Subsystems V: Structural and Thermal," in *Space Mission Engineering: The New SMAD* (Space Technology Library, 2011), p. 879.

[31] Several international protocols have been implemented through the Consultative Committee on Space Data Services. See https://public.ccsds.org/default.aspx; ibid., p. 894.

Three-quarters of the revenue stems from geolocation and navigation equipment, such as Global Positioning System (GPS) receivers.[32]

Space-Related Products and Services

Global commercial space products and services generated $126 billion in revenue in 2015. The space component of some of these products and services is not always well understood by the users, who may take some of them—such as GPS—for granted. The major commercial categories include the following:

- *Communications systems* relay radio and television signals sent from a point on the ground to a satellite and then to another ground point. These signals may carry such content as television programs, in-flight calls from airplane passengers, and some smartphone data. Satellite television alone accounts for almost a third of all space-related commercial activity.[33]
- *Earth observation* provides environmental monitoring of oceans, forests, deserts, wildlife habitats, and natural disasters.
- *Global atmospheric monitoring* includes data collected for meteorological use to help predict weather patterns, hurricanes, and El Niño. It also measures soil water content to assist in prediction of droughts and floods.
- *Transportation* uses provide geolocation services to delivery trucks and ride-sharing services and their passengers.
- *Safety enhancement* provides data to first responders at oil spills and forest fires and prevents train collisions with geolocation services.

---

[32] Space Foundation, *The Space Report*, 2016, pp. 17-19.

[33] The U.S. satellite industry, comprising satellite manufacturing, telecomm services (including television, mobile data, remote sensing), launch vehicles, and ground equipment, is estimated as having $208 billion in revenue in 2015. See Satellite Industry Association, *State of the Satellite Industry Report*, June 2016, p. 5.

Satellites also have noncommercial, national security purposes, including detecting the launch of missiles; detecting nuclear explosions in the ground or atmosphere to monitor nuclear treaty compliance; and providing global jam-resistant communications for strategic and tactical forces during a conflict.

INSURANCE

Space launches are risky: three of the 86 commercial launches attempted in the United States in 2015 failed, resulting in the destruction of launch vehicles and costly payloads. Given the potential losses, insurance coverage is likely to play an important role in the development of a commercial space industry. Worldwide space insurance premiums were more than $700 million in 2015, and insured losses exceeded $600 million in 2014.[34] Major insurers such as American International Group, Munich Re, and Allianz compete in the market.

Nonetheless, insurance on space-related risks appears to have limitations. In September 2016, a rocket owned by SpaceX exploded while on the launchpad at Cape Canaveral, Florida, destroying the satellite it was preparing to launch. The Falcon 9 rocket was reportedly not insured. The owner of the satellite it was attempting to launch held $300 million of coverage. However, the policy[35] may not be applicable because the rocket explosion took place during a prelaunch test and not during the actual launch.[36]

---

[34] Ibid., p. 20.

[35] The satellite destroyed in the SpaceX explosion was built by Israeli-based Space Communication for Facebook and Eutelsat; it was designed to provide Internet service to sub-Saharan Africa. Dana Hull and Julie Johnsson, "SpaceX Explosion Draws Attention to Insurance for Rocket Science," *Insurance Journal*, September 2, 2016.

[36] Ian Salisbury, "Elon Musk's SpaceX Rocket Could Have Been Insured, But Wasn't," *Time*, September 2, 2016.

NEW ENTRANTS CHANGE THE INDUSTRY

Three developments are changing the shape of the commercial space industry: a shift in government space activities toward the use of commercial services, an increase in private financing, and an increase in the launch of small satellites. These changes are supporting the development of new entrants with new launch products.

U.S. Government Redefines Procurement

Whereas NASA once owned the spacecraft produced for it by suppliers like Boeing, it now is in some cases transferring risk by contracting out for services, leaving the ownership of the launch vehicles to commercial entities. Its choice of traditional procurement or a more commercially oriented approach (Table 1) depends on the program mission.[37] With the commercially oriented approach, NASA agrees on a fixed price for the services a contractor is to provide, rather than using a cost-plus methodology that reimburses the contractor's allowable expenses and adds an additional payment to ensure a profit.

One example of the new approach is procurement of transportation of astronauts to the ISS. NASA traditionally would have provided detailed descriptions of each step in development of a launch vehicle and funded it as well. Now, however, more project development is left to the contractor, which shares the cost.[38] The commercial partner is expected to meet project milestones. If it does, it is paid and the project moves to its next stage.

---

[37] NASA has worked with private companies almost from its inception, developing the first civil communications satellites in cooperation with AT&T, Radio Corporation of America (RCA), Hughes Aircraft, and others. Roger Cochetti, *Mobile Satellite Communications Handbook*, 2nd ed. (Hoboken, NJ: John Wiley & Sons, Inc., 2015), p. 3.

[38] NASA's Commercial Orbital Transportation Services (COTS) is an example of a public-private partnership using the approaches in Table 1. SpaceX and Orbital LEO transportation systems were developed jointly: SpaceX invested 53% and the U.S. government invested 47% for the development and demonstration of a commercial transportation system; Orbital invested 58% and the U.S. government invested the remaining 42%. NASA, *Commercial*

Table 1. Distinct Approaches for Distinct Missions at NASA

Program Characteristics	Traditional Approach	Commercial-Oriented Approach
Owner	NASA	Industry
Contract type	Cost-plus	Fixed price
Contract management	Prime contractor	Public-private partnership
Customers	NASA	Government and nongovernment
Funding for capability development	NASA procures capability	NASA provides "milestone" payments as agreed goals are reached
NASA's role in capability development	NASA defines "what" and "how"	Industry defines "how" and NASA approves
Requirements definition	NASA defines detailed requirements	NASA defines only higher-level requirements
Cost structure	NASA incurs total cost	NASA and industry share costs

Source: NASA, Human Exploration and Operations Mission Directorate.

During the final years of the space shuttle program, NASA encouraged and funded commercial providers to develop systems that could transport crew and cargo to the ISS. The first of these was the SpaceX Dragon capsule and Falcon 9 rocket, which has ferried supplies to the ISS since 2012. Last year, four ISS cargo launches were conducted by Orbital ATK and three more by SpaceX. (Boeing and SpaceX are each developing ISS crew transportation capabilities for first use in the next few years).

New entrants are changing the economics of launches by reuse of rocket boosters. SpaceX's entry into the launch market provides NASA and DOD with new options, and it is also cutting into the international launch market formerly dominated by foreign providers. SpaceX has had several successful launches and relandings. Blue Origin has to date launched several prototypical rockets that have returned to their launchpads.

Orbital Transportation Services: A New Era in Spaceflight, February 2014, p. 95, https://www.nasa.gov/sites/default/files/files/ SP-2014-617.pdf.

Private Financing Increases

Space-related industries[39] are taking advantage of a range of private financing options, including venture capital, debt financing, and acquisition, receiving more than $13 billion in such investments between 2000 and 2015. According to Tauri Group, a space industry research organization, this trend has accelerated over the past five years, with 2015 recording $2.3 billion in space-related capital investment, a record level for one year.[40] Several wealthy business owners—some with ties to Silicon Valley—and corporations are also investing in space-related companies,[41] as are banks, including the U.S. Export-Import Bank.[42] These combined capital sources have spurred the establishment of new companies: In the last five years, an average of eight new space ventures were established annually, almost triple the level of such firm creation in the early 2000s.[43] According to Tauri Group, investors find space attractive because launch vehicle costs are forecast to drop with the use of new types of launch hardware, including reusable rockets, and because they anticipate the development of new products and services,[44] such as space tourism and new satellite sources for collecting and analyzing data obtained in space.

---

[39] Including satellite, launch vehicle, and ground equipment manufacturing; satellite-based services such as television, radio, and broadband; and data collection and analytical services. Tauri Group, *Start-Up Space: Rising Investment in Commercial Space Ventures*, January 2016, p. 2.

[40] Of the $2.3 billion invested in space-related companies in 2015, $1.8 billion was venture capital. Ibid., p. iv.

[41] So-called space billionaires Jeff Bezos, Elon Musk, and Richard Branson have formed start-up space ventures. Large companies such as Google, Qualcomm, AT&T, Coca-Cola, and Monsanto are also investing. Ibid., p. 9.

[42] Ex-Im Bank has provided debt funding for satellite systems. Ibid., p. 10.

[43] The availability of private funding can affect NASA's commercial space resources. In 2006, NASA chose SpaceX and Rocketplane Kistler (RpK) to provide future space transportation services to the ISS. In 2007, NASA terminated its contract with RpK because the company was unable to raise enough private funding to continue its vehicle development. NASA, *Commercial Orbital Transportation Services*, NASA/SP-2014-617, February 2014, p. v.

[44] Long-established aerospace firms are not part of the Tauri Group analysis, even if they obtain private equity investments; new space ventures and capital directed to them are the focus of that analysis. The Tauri Group, *Start-Up Space: Rising Investment in Commercial Space Ventures*, January 2016, p. 3.

Growing Deployment of Small Satellites

The development of small satellites[45] for Earth imaging and establishing space-based Internet networks from LEO is now possible because satellite components have been miniaturized and standardized. Groups of small satellites are referred to as constellations; Planet Labs, for example, has a constellation of 36 small satellites in orbit, with customers paying for the images it can capture at less distance from Earth than is possible with larger satellites in higher orbits.[46] Some observers argue that demand for data may be driving the market for small satellites, as much as the new technologies.[47]

Small satellites allow access to space by researchers, companies, and governments that cannot afford larger spacecraft. Because small satellites can travel as a secondary payload on many launch vehicles, launch costs may be only a few million dollars per satellite, although multiple small satellites may be required for many purposes.[48]

The small satellite market is addressed by a number of startup firms that hope to succeed in providing broadband, remote imaging, or communication services, such as Firefly Space Systems in Texas, Rocket Lab in California, and OneWeb in Virginia. While small satellites are normally launched on rockets with other, larger payloads, Virgin Galactic is proposing a new type of launcher: a rocket attached to the wing of a modified commercial 747 jet will launch a payload into orbit when the plane reaches an altitude of 35,000 feet. Virgin Galactic believes this form

---

[45] Small satellites are categorized as picosatellites (under 1 kilogram [kg]), nanosatellites (1-10 kg), and microsatellites (10-100 kg). CubeSats, which generally weigh less than 1.3 kg, are in the shape of a cube that can fit in a person's hand. In the last 50 years, 38 picosatellites, 680 nanosatellites, and 860 microsatellites have been launched worldwide. Henry Helvajian and Siegfried W. Janson, eds., *Small Satellites: Past, Present and Future* (El Segundo, CA: Aerospace Press, 2009).

[46] Clay Dillow, "Here's why small satellites are so big right now," *Fortune*, August 4, 2015.

[47] Prateep Basu, *'Big Data' Leap in EO Markets*, Northern Sky Research, October 1, 2016, http://www.nsr.com/newsresources/the-bottom-line/big-data-leap-in-eo-markets/.

[48] Aerospace Press, "Small Satellites: Past, Present, and Future," press release, June 1, 2008, http://www.aerospace.org/publications/aerospace-books/small-satellites-past-present-and-future.

of launch vehicle, using smaller rockets, will significantly reduce the cost of putting small satellites into orbit.[49]

NASA is also utilizing small so-called CubeSats to address scientific questions and broaden the involvement of students and researchers. In 2015, 42 small satellites, primarily with commercial functions, were launched as cargo to the ISS for later deployment from there.[50] NASA has announced that it will help develop new CubeSat technologies and will launch six small Earth-observing satellite missions.[51]

COMMERCIAL SPACE WORKFORCE DECLINES

The space industry workforce includes employees in private-sector firms as well as those working at NASA, DOD, and other government agencies. The increasing public attention given to the commercial space industry belies a dichotomy: the commercial and civil space workforces are declining, while national security workforces remain steady.

Commercial employment in the space industry peaked in 2006 at about 267,000 employees and has declined steadily since then.[52] The termination of NASA's space shuttle program[53] in 2011 resulted in the loss of many private-sector jobs. Other major reasons for the decline in space industry employment may be many skilled workers reaching retirement age and difficulties in recruiting young talent.[54]

---

[49] Samantha Masunaga, "Small satellites are back, with down-to-earth expectations," *Los Angeles Times*, May 27, 2016.

[50] Federal Aviation Administration, *The Annual Compendium of Commercial Space Transportation*, January 2016, p. 36.

[51] NASA, "New CubeSats to Test Earth Science Tech in Space," press release, February 22, 2016, http://www.nasa.gov/press-release/nasa-selects-instruments-to-study-air-pollution-tropical-cyclones; NASA, "NASA Small Satellites Set to Take a Fresh Look at Earth," press release, November 7, 2016, http://www.nasa.gov/pressrelease/nasa-small-satellites-set-to-take-a-fresh-look-at-earth. The new satellites will reach orbit on an Orbital ATK Pegasus XL rocket.

[52] Bureau of Labor Statistics, *Quarterly Census of Employment and Wages (QCEW)*.

[53] The United Space Alliance was formed by Boeing and Lockheed Martin in 1996 as the major contractor for the shuttle and the ISS. See http://www.unitedspacealliance.com.

[54] U.S. Department of Commerce, Bureau of Industry and Security, *U.S. Space Industry "Deep Dive" Assessment: Employment in the U.S. Space Industrial Base*, September 2014, p. 13.

Table 2. Space Industry Employment (2005-20 15)

Industry Sector	Employment in Sector		Percent Change
	2005	2015	
Search, Detection, and Navigation Instruments (*334511*)[a]	155,492	124,578	-19.9%
Guided Missile and Space Vehicle Manufacturing (336414)[b]	53,316	55,825	+ 4.7
Guided Missile and Space Vehicle Propulsion Unit and Parts Manufacturing (*336415*)[c]	13,115	9,729	-25.8
Other Missile and Space Vehicle Manufacturing (336419)[d]	7,423	5,309	-28.4
Satellite Telecommunications (*517410*)[e]	16,349	8,633	-47.2
Space Research and Technology (927110)[f]	18,371	17,293	-5.9
TOTAL	264,066	221,367	-16.2%

Source: Bureau of Labor Statistics, Quarterly Census of Employment and Wages (QCEW).

Notes: Industry categories are based on the North American Industry Classification System (NAICS); NAICS codes are shown italicized above.

[a] Manufacturing of search, detection, navigation, guidance, aeronautical, and nautical systems and instruments.

[b] Manufacturing of complete guided missiles and space vehicles and/or developing and making prototypes of such missiles and vehicles.

[c] Manufacturing of guided missile and/or space vehicle propulsion units and parts and/or developing and making prototypes of such units and parts.

[d] Manufacturing of guided missiles and space vehicle parts and auxiliary equipment (except guided missile and space vehicle propulsion units and parts) and/or developing and making prototypes of guided missile and space vehicle parts and auxiliary equipment.

[e] Telecommunications services provided to other establishments in the telecommunications and broadcasting industries by forwarding and receiving communications signals via system of satellites or reselling satellite communications.

[f] Government establishments primarily engaged in the administration and operations of spaceflights, space research, and space exploration, including government establishments operating spaceflight centers.

Jobs in this industry generally require advanced skills, such as engineering, differentiating them from the average U.S. manufacturing or service-industry job. The Space Foundation has identified six industry sectors that collectively approximate the space industry. Bureau of Labor Statistics (BLS) data show that employment in these six sectors dropped by 16% in the 10 years from 2005 to 2015 (Table 2).[55]

---

[55] These six categories also include jobs that are not in the commercial space industry. For example, NAICS 334511 also includes workers in broader aerospace and nautical industries, so some of the decline in that category could have been in those industries, not

These data do not show the full scope of the space industry workforce. The categories listed in Table 2 include both space and nonspace activity,[56] while employees in other sectors who are working on space-related tasks are excluded because most of the employment in the sector is not related to space.[57]

An alternative approach to assessing the size and breadth of the space industry workforce is the 2012 space industry report by the U.S. Department of Commerce (DOC), based on a survey of 3,780 respondents in 16 different manufacturing and service segments of the space economy, including spacecraft and launch vehicles, communications systems, ground systems, electronic equipment, and software.[58] This survey, conducted once and covering the years from 2009 to 2012, shows a somewhat larger workforce of 254,179 employees in 611 commercial companies that viewed themselves as dependent on federal space programs.[59]

In addition to commercial employment, DOC identified 13,897 space-related jobs in the U.S. government, 19,630 in nonprofit organizations, and 60,533 at universities, for a total of 348,239 workers dependent on space-related programs. The report shows shrinkage of nearly 7% in employment at companies dependent on commercial space programs from 2009 to 2012.[60]

Wages in the commercial space industry are higher than average U.S. wages. The average space industry salary of $111,000 (in 2014)[61] is double what the average private-sector employee receives.[62] Average salaries vary

---

commercial space. Data from the Bureau of Labor Statistics (BLS) do not go into that greater level of detail.

[56] For example, NAICS 334511 includes manufacture of space guidance systems, but also the manufacture of nonspace equipment such as aircraft airspeed measurement instruments.

[57] For example, NAICS Surveying and Mapping Services (54137) includes imagery obtained by satellites, but the larger part of the category is based on aerial photography.

[58] U.S. Department of Commerce, Bureau of Industry and Security, *U.S. Space Industry "Deep Dive" Assessment: Employment in the U.S. Space Industrial Base*, September 2014, pp. 18-19.

[59] Employment estimate is for 2012, with 611 of the 3,780 respondents saying they were dependent on federal space-related programs for their viability. Each company determined its own standard of dependency. Ibid.

[60] Ibid.

[61] Space Foundation, *The Space Report*, 2016, p. 65.

[62] In 2015, the average U.S. private sector job paid $52,876, and the average U.S. manufacturing job paid $64,305. Bureau of Labor Statistics, *Quarterly Census of Employment and Wages*.

within the six industry categories cited in Table 2, from about $92,000 in missile and vehicle propulsion (NAICS 336415) to more than $124,000 in missile and vehicle manufacturing (NAICS 336414).[63]

POLICY ISSUES FOR CONGRESS

Three overarching issues will affect the development of commercial space in the future: how the industry is regulated by diverse federal agencies, the effects of new export control laws and regulations that seek to increase U.S. space industry competitiveness, and the allocation of spectrum for satellite use.

Regulating and Managing Commercial Space

The commercial space industry is governed by federal agencies with diverse regulatory interests. The major federal agencies with commercial space policy responsibilities are as follows:

- *FAA's Office of Commercial Space Transportation* regulates launches and reentries of space vehicles, as well as the U.S. launch and reentry sites, including rerouting aircraft that may interfere with a nearby rocket launch.[64]
- The *Department of Commerce* has two major space industry interests. The National Oceanic and Atmospheric Administration (NOAA) licenses commercial imaging satellites and utilizes commercial space capabilities for improved weather forecasting and environmental data collection,[65] and also oversees the Office of Space Commerce, which promotes the U.S. commercial space

---

63 Space Foundation, *The Space Report*, 2016, p. 66.

64 https://www.faa.gov/about/office_org/headquarters_offices/ast/regulations.

65 http://www.noaanews.noaa.gov/stories2016/images/NOAA%20Commercial%20Space%20 Policy.pdf.

industry's economic growth and technological advancement.[66] DOC's Bureau of Industry and Security (BIS) administers export controls and licensing for strategic technologies, including space industry components.[67]

- *NASA* provides infrastructure and operations support and encourages private-sector investment in its launches and other activities through its Commercial Crew and Cargo Programs.[68]
- *Department of Defense* utilizes commercial space systems and technologies for national security purposes. Through the Air Force, it provides infrastructure, operations support, and safety oversight for government and commercial launches at its launch sites.[69] Additionally, through the Defense Technology Security Administration (DTSA), it reviews and comments on applications for export licenses, in conjunction with Department of State review.[70]
- *Department of State* (DOS) is responsible for the export and temporary import of defense articles and services—including some commercial space components— through the International Traffic in Arms Regulations (ITAR).[71]
- *Federal Communications Commission* licenses commercial satellite radio frequencies and determines placement of satellites in geostationary orbit.[72]

The current regulatory and management structure may require changes if the commercial space industry grows. Private firms may take on more responsibilities for launching government satellites (including military

[66] http://www.space.commerce.gov/about/mission.
[67] Most space technologies are covered by U.S. export control laws and regulations. Export Administration Regulations (EAR) cover dual-use products that could be used for either commercial or military purposes. https://www.bis.doc.gov/index.php/about-bis.
[68] https://www.nasa.gov/exploration/commercial/index.html.
[69] http://archive.defense.gov/home/features/2011/0111_nsss/docs/Fact%20Sheet%20DoD%20Space%20Policy.pdf.
[70] http://www.dtsa.mil/SitePages/assessing-and-managing-risk/space-launch-technology.aspx.
[71] http://www.pmddtc.state.gov/regulations_laws/itar.html.
[72] https://www.fcc.gov/licensing-databases/licensing and https://www.fcc.gov/approved-space-station-list.

satellites), offer space flights for tourists, undertake mining operations on asteroids and the moon, and attempt to explore and even settle Mars. This could lead to greater commercial demand for FAA to process licenses, permits, and safety reviews. Congress may need to set additional legal parameters to establish a legal basis for regulation of passenger safety in space tourism and space traffic management. In addition, the current legal provision for liability risk-sharing between launch companies and the U.S. government may require changes as the industry matures and more private individuals become passengers or crew.

A better system for integrating commercial space launches into the air traffic control system is a particular concern for FAA. Currently, when a launch or reentry is planned, FAA closes airspace near the site because equipment onboard aircraft often cannot track rockets and spacecraft moving at very high speeds. Since launch times may vary from initial plans, FAA may close the airspace for a long period, causing both commercial and military aircraft to be delayed or rerouted. The FAA alert system utilizes email and phone calls to commercial and military airspace users. As the number of rocket launches increases, FAA may face pressure to reduce the length of air space closures. The agency plans to use an automated system to deliver alerts to aircraft pilots to minimize air traffic impact.[73]

In the 114th Congress, H.R. 4945 has been proposed to change federal oversight of commercial space. In addition to making organizational changes within DOD, the legislation would direct NASA to develop a 20-year plan to land astronauts on Mars, and would require the Department of Transportation to establish an Office of Commercial Space Transportation and an Office of Spaceports and also to designate a lead government agency for space traffic management. DOC would be directed to develop a plan to coordinate space-related economic and regulatory activities, and DOS would be required to begin developing an international traffic management regime.

---

[73] Michael Lopez-Alegria, Former Astronaut and Vice Chairman, Commercial Space Transportation Advisory Committee (COMSTAC), *FAA Oversight of Commercial Space Transportation*, Subcommittee on Aviation, House Committee on Transportation and Infrastructure, June 22, 2016, pp. 5-6.

Export Controls

Spacecraft, ground stations, and some unique components are regulated by the U.S. export control system. They are considered dual-use items, as even those developed primarily for commercial or civil purposes have possible military applications. During the Cold War era, these space industry products were regulated exclusively by DOS[74] and were considered munitions, as many of them were then designed specifically for military purposes. Although the regulation of these dual-use space products was transferred to DOC[75] in the 1990s, Congress returned regulatory authority to DOS in 1998 after some satellite designs were improperly transferred to China.[76]

DOS's administration of export controls came under strong criticism from manufacturers of space-related equipment. The satellite industry asserted that the long licensing process led to a loss of sales abroad, and the Aerospace Industries Association (AIA) claimed that the U.S. share of the global commercial satellite market fell from 63% before export controls were transferred to DOS to 30%.[77] A DOC survey of manufacturers found that many believed that the DOS export controls "eroded U.S. competitiveness in the international space market."[78] In 2013, Congress transferred export control responsibilities for 80% of satellites and related items back to DOC regulation for sales to most countries.[79] AIA and others are calling for controls on some commercial items that are

---

[74] Department of State licensing program is through the International Traffic in Arms Regulation (ITAR).

[75] The Department of Commerce administers Export Administration Regulations (EAR) through the Bureau of Industry and Security.

[76] Before 1990, commercial communications satellites and related items were on the State Department's U.S. Munitions List (USML). For more information on the U.S. export control system and regulation of satellites and related equipment, see CRS Report R41916, *The U.S. Export Control System and the President's Reform Initiative*, by Ian F. Fergusson and Paul K. Kerr, p. 19.

[77] David Melcher, "Tick Tock," *Space News*, October 10, 2016, http://www.spacenewsmag.com.

[78] Bureau of Industry and Security, Department of Commerce, *U.S. Space Industry "Deep Dive" Assessment: Impact of U.S. Export Controls on the Space Industrial Base*, February 2014, p. 14.

[79] National Defense Authorization Act of 2013, P.L. 112-239; satellites exports remain prohibited to China, North Korea, and countries designated as state sponsors of terrorism (currently Iran, Sudan, and Syria).

still under DOS, such as apertures on electro-optical satellites used for remote sensing, integrated propulsion systems, and plasma thrusters, to be transferred to DOC.[80]

AIA also has raised concerns about the launch vehicle restrictions in the international Missile Technology Control Regime (MTCR), which regulates missile proliferation. As space tourism becomes a possibility, some MTCR rules—as incorporated into U.S. export controls—could affect the development of this commercial activity, including limitations on the type of rocket that could be used to boost space tourists into suborbital space. In addition, a space tourism rocket that is launched in one country and descends in another country could be considered a missile export subject to strict MTCR controls.[81]

Spectrum Allocation

The satellite industry[82] is concerned that sharing of certain bands for future wireless 5G use— some of which are currently used solely for satellite transmissions—could jeopardize the reliability and cost-effectiveness of their service.[83]

For decades, satellites have communicated on dedicated frequency bands assigned by the International Telecommunications Union (ITU). The specific band used by a satellite depends on its purpose.[84] For example,

---

80 Aerospace Industries Association and Satellite Industry Association, industry analysis submitted to Department of State, *Satellite Export Control Priorities*, 2015.

81 CRS conversation with staff of Aerospace Industries Association, November 2, 2016, and Peter de Selding, "U.S. Satellite Export Regs Remain a Frustration for European Industry," *Space News*, February 7, 2014, http://spacenews.com/39414us-satellite-export-regs-remain-a-frustration-for-european-industry/.

82 Letter from Tom Stroup, President, Satellite Industry Association, to Marlene Dortch, Secretary, Federal Communications Commission, October 5, 2012; Monica Alleven, "O3b throws water on idea of near-term solution for 28 GHz sharing," *Fierce Wireless*, April 29, 2016.

83 The commercial space industry is not alone in arguing against spectrum sharing; other industries, including nascent autonomous motor vehicles, railroads, and federal weather satellite services, have made similar arguments to the FCC.

84 Satellite frequency bands range from 1,467 gigahertz (GHz) to 30 GHz. EMEA Satellite Operators Association (ESOA), *Satellite Spectrum*, viewed November 4, 2016, https://www.esoa.net/spectrum/satellite-spectrum.asp.

lower frequencies—the L-, S-, and C-bands—are not affected by heavy rainfall and are therefore used in tropical regions; they can also simultaneously serve large areas of the globe. The Ku and Ka bands are used for television broadcasting and data services. Satellites are built to transmit a specific frequency and cannot be reprogrammed after launch.[85]

The Obama Administration directed the FCC to identify spectrum that could be used to expand Wi-Fi services in the future. To address that goal, FCC and the National Telecommunications and Information Administration (NTIA) collaborated in identifying 500 MHz of spectrum suitable for wireless broadband use, including future domestic 5G terrestrial mobile providers.[86] Some of this spectrum was obtained by sharing frequencies in the 28 GHz band previously dedicated to satellite transmissions.[87]

The satellite industry expressed concern over sharing spectrum, particularly the 28 GHz band, contending that current satellite services could be undermined without a deeper understanding of how future 5G services will be used:

> [S]atellites are providing vital services to all Americans using spectrum bands above 24 GHz. Satellites "distribute point-to-multipoint video and other high bandwidth services more efficiently and more cost effectively than any other technology." In addition, high-throughput satellites are bringing competitive broadband services to all of the United States. Satellites also provide advanced services to ships, aircrafts and motor vehicles.... Consideration of repurposing spectrum access from existing users with a supporting record would result in the loss of critical

---

[85] Ibid.

[86] Department of Commerce, *Plan and Timetable to Make Available 500 Megahertz of Spectrum for Wireless Broadband*, October 2010, p. ii, https://www.ntia.doc.gov/files/ntia/publications/tenyearplan_11152010.pdf.

[87] Federal Communications Commission, "Use of Spectrum Bands Above 24 GHz," 81 *Federal Register* 79894-79945, November 14, 2016, https://www.gpo.gov/fdsys/pkg/FR-2016-11-14/pdf/2016-25765.pdf.

services to U.S. consumers, and to vital enterprise users such as first responders and the U.S. military.[88]

The FCC contends that its approach in its July 2016 ruling struck a balance between wireless services and satellite operations and that the spectrum sharing provided in the decision would "ensure that diverse users—including federal and non-federal, satellite and terrestrial, and fixed and mobile—can co-exist and expand."[89] The FCC also noted that there had been "dueling studies" from affected industries and said it would continue to study the issue and make adjustments if necessary.[90]

There has been congressional interest in the FCC's activities with regard to satellite spectrum. H.R. 4945 (114th Congress) would direct the FCC to preserve primary electromagnetic access in the 27.5-28.35 GHz band for satellite operators.[91] The FCC is also evaluating a proposal by Ligado Networks, a satellite-communications company developing a network to support 5G service, for sharing the 1675-1680 band with NOAA. DOC, the Air Force, and the Aerospace Industries Association have expressed concern that this sharing arrangement might interfere with GPS signals and the emergency response, homeland security, and aviation safety sectors that rely on them.[92]

---

[88] Letter from Tom Stroup, President, Satellite Industry Association, to Federal Communications Commission, February 18, 2015, http://www.sia.org/wp-content/uploads/2015/02/SIA_Reply_Comments_Spectrum_Frontiers_2015_02_18.pdf.

[89] Federal Communications Commission, "FCC Takes Steps to Facilitate Mobile Broadband and Next Generation Wireless Technologies in Spectrum Above 24 GHz," press release, July 14, 2016, https://apps.fcc.gov/edocs_public/attachmatch/DOC-340301A1.pdf. GN Docket No. 14-177.

[90] John C. Tanner, "What the FCC's 5G Spectrum Plan Means for Ka-band Satellite," telecomasia.net, July 20, 2016.

[91] H.R. 4945, American Space Renaissance Act. The comprehensive legislation was referred to several committees.

[92] The Commander of the Air Force Space Command mentioned his concern during a March 15, 2016, hearing by the House Subcommittee on Strategic Forces. Letter from Ali Bahrami, Vice President, Civil Aviation, Aerospace Industries Association, to Federal Communications Commission, June 21, 2016, https://ecfsapi.fcc.gov/file/10621054184363/Ligado%20-%20IB%20Docket%20No%20%20RM-11681_AIA-FINAL.pdf; Dee Ann Divis, "FCC Opens GPS-Adjacent Ligado Proposal for Comment," *Inside GNSS News*, May 1, 2016, p. http://www.insidegnss.com/ node/4933.

APPENDIX A. SATELLITE SIZE COMPARISON

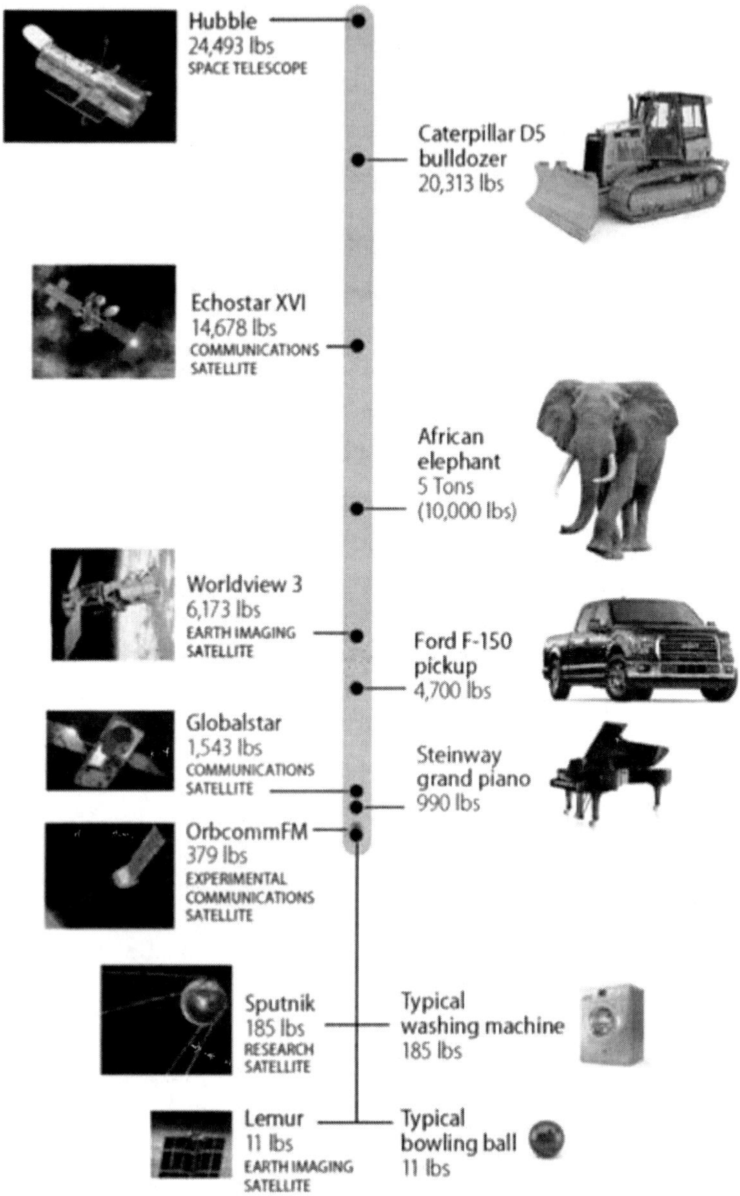

Source: CRS.

Figure A-1. Types of Satellites Based on Mass, Compared to Earth-Bound Objects.

In: Commercial Crew Program … ISBN: 978-1-53614-980-7
Editor: Graeme N. Lutz

Chapter 4

NASA: Preliminary Observations on the Management of Space Telescopes*

Cristina T. Chaplain

Why GAO Did This Study

Acquisition management has been a long-standing challenge at NASA, although GAO has reported on improvements the agency has made in recent years. Three space telescope projects are the key enablers for NASA to achieve its astrophysics' science goals, which include seeking to understand the universe. In its fiscal year 2018 budget request, NASA asked for about $697 million for these three projects, which represents over 50 percent of NASA's budget for its astrophysics' major projects. In total, these projects represent an expected investment of at least $12.4 billion.

This statement reflects preliminary observations on (1) the current status and cost of NASA's major telescope projects and (2) lessons learned

* This is an edited, reformatted and augmented version of a United States Government Accountability Office Testimony before the Subcommittee on Space, Committee on Science, Space, and Technology, House of Representatives, Accessible Version, Publication No. GAO-18-277T, dated December 6, 2017.

that can be applied to NASA's management of its telescope projects. This statement is based on ongoing work on JWST and ongoing work on the status of NASA's major projects. Both reports are planned to be published in Spring 2018. This statement is also based on past GAO reports on JWST and NASA's acquisitions of major projects, and NASA input.

What GAO Recommends

GAO is not making any recommendations in this statement, but has made recommendations in prior reports to strengthen NASA's acquisition management of its major projects. NASA has generally agreed with GAO's recommendations and taken steps to implement them.

What GAO Found

The National Aeronautics and Space Administration's (NASA) current portfolio of major space telescopes includes three projects that vary in cost, complexity, and phase of the acquisition life cycle.

GAO's ongoing work indicates that these projects are each making progress in line with their phase of the acquisition cycle but also face some challenges. For example, the current launch date for the James Webb Space Telescope (JWST) project reflects a 57-60-month delay from the project's original schedule. GAO's preliminary observations indicate this project still has significant integration and testing to complete, with very little schedule reserve remaining to account for delays. Therefore, additional delays beyond the delay of up to 8 months recently announced are likely, and funding available under the $8 billion Congressional cost cap for formulation and development may be inadequate.

There are a number of lessons learned from its acquisitions that NASA could consider to increase the likelihood of successful outcomes for its telescope projects, as well as for its larger portfolio of projects, such as its human spaceflight projects.

Table 1. Current Phase, Cost, and Schedule Status of National Aeronautics and Space Administration's (NASA) Major Space Telescope Projects

Project	**Phase**	**Preliminary Cost Estimate (dollars in millions)**		**Preliminary Schedule**	
Wide-Field Infrared	Formulation	3,200-3,800		2024-2026	
Survey Telescope	(early phase)				
		Current Cost Estimate (dollars in millions)	Change from Baseline (dollars in millions)	Target launch date	Change from baseline (months)
Transiting Exoplanet Survey Satellite	Implementation (building, launching, and operating)	336.7	-41.7[a]	March 2018	-3
James Webb Space Telescope	Implementation (building, launching, and operating)	8,825.4	3,861.8	March-June 2019	57-60

Source: GAO analysis of National Aeronautics and Space Administration data. | GAO-18-277T.

[a]The dollar change reflects a decrease of $26.7 million after launch vehicle selection in 2014 due to the reduction in planned costs and a decision by NASA in August 2017 to reallocate $15 million of the project's headquarters-held reserves to the Wide-Field Infrared Survey Telescope project.

For example, twice in the history of the JWST program, independent reviews found that the program was not holding adequate cost and schedule reserves. GAO has found that NASA has not applied this lesson learned to all of its large projects, and similar outcomes to JWST have started to emerge. For example, NASA did not incorporate this lesson with its human spaceflight programs. In July 2016 and April 2017, GAO found that these programs were holding inadequate levels of cost and schedule reserves to cover unexpected cost increases or delays. In April 2017, GAO recommended that NASA reassess the date of the programs' first test flight. NASA concurred and, in November 2017, announced a launch delay of up to 19 months.

* * *

Chairman Babin, Ranking Member Bera, and Members of the Subcommittee:

I am pleased to be here today to discuss the National Aeronautics and Space Administration's (NASA) management of its astrophysics' space telescope projects. These telescopes are the key enablers for the agency to achieve its astrophysics' science goals, which include seeking to understand the universe and our place in it. These major space telescope projects—projects with a life cycle cost greater than $250 million—include:

- the James Webb Space Telescope (JWST), which is designed to help understand the origin and destiny of the universe, the creation and evolution of the first stars and galaxies, and the formation of stars and planetary systems;
- the Transiting Exoplanet Survey Satellite (TESS), whose mission goal is to discover exoplanets—or planets in other solar systems—during transit, the time when the planet's orbit carries it in front of its star as viewed from Earth; and
- the Wide-Field Infrared Survey Telescope (WFIRST), which is designed to perform wide-field imaging and survey of the near-infrared sky to answer questions about the structure and evolution of the universe and expand our knowledge of planets beyond our solar system.

In its fiscal year 2018 budget request, NASA asked for about $697 million for these three projects, which represents over 50 percent of

NASA's budget for its astrophysics' projects.[1] In total, these projects represent an expected investment of at least $12.4 billion. As such, while it is important for NASA to continually stretch technological boundaries to further scientific research, it is also important to manage these projects prudently, with clear accountability and oversight for taxpayer dollars.

For over two decades, acquisition management has been a long-standing challenge at NASA, although we have reported on improvements the agency has made in recent years.[2] We first designated NASA's acquisition management as a high-risk area in 1990 in view of NASA's history of persistent cost growth and schedule slippage in the majority of its major systems. Our work has identified a number of causal factors related to these issues, including poor cost estimating and underestimation of risks associated with the development of its major systems. We have also identified a set of best practices that can help agencies manage development risks. NASA leadership has made concerted efforts to address these causal factors. In our February 2017 High Risk Update, however, we found that more needs to be done with respect to anticipating and mitigating risks—especially with regard to large programs, estimating and forecasting costs for its largest projects, and implementing management tools.[3]

My statement today provides our preliminary observations on (1) the current status and cost of NASA's major telescope projects and (2) lessons learned that can be applied to NASA's management of its telescope projects. This statement is based on our ongoing work for this committee and others on the JWST project and our annual review of the status of all of NASA's major projects, as well as our February 2017 High- Risk Update and other past reports.[4] To assess the cost and schedule

[1] According to NASA officials, NASA historically spends 50-70% of its astrophysics budget on developing new missions.

[2] For example, see GAO, *NASA: Assessments of Major Projects*, GAO-17-303SP (Washington, D.C.: May 16, 2017).

[3] GAO, *High-Risk Series: Progress on Many High-Risk Areas, While Substantial Efforts Needed on Others*, GAO-17-317 (Washington, D.C.: Feb. 15, 2017).

[4] GAO, *James Webb Space Telescope: Actions Needed to Improve Cost Estimate and Oversight of Test and Integration,* GAO-13-4 (Washington, D.C.: Dec. 3, 2012); GAO, *NASA: Assessments of Selected Large-Scale Projects*, GAO-13-276SP (Washington, D.C.: Apr. 17,

performance of these projects, we collected information on these areas from projects using a data collection instrument, analyzed projects' monthly status reports, interviewed NASA project and headquarters officials, and reviewed project documentation. For JWST and TESS, which are in the implementation phase, we compared current cost and schedule estimates to their original cost and schedule baselines. To identify lessons learned that can be applied to NASA's management of its telescope projects, we examined NASA's efforts to address issues identified in our prior JWST work, such as the quality of the cost and schedule risk analyses, and our February 2017 High-Risk Update.[5]

We are conducting the work on which this statement is based in accordance with generally accepted government auditing standards. We plan to issue a final report on our annual review of the JWST program, as well as our annual assessment of NASA's major projects, in Spring 2018. NASA provided us technical comments on information that is included in this statement on the telescope projects, which we incorporated as appropriate.

BACKGROUND

NASA's mission is to drive advances in science, technology, aeronautics, and space exploration, and contribute to education, innovation, our country's economic vitality, and the stewardship of the Earth. To accomplish this mission, NASA establishes programs and projects that rely on complex instruments and spacecraft. NASA's portfolio of major projects ranges from space satellites equipped with advanced sensors to study the Earth to a telescope intended to explore the universe to spacecraft to transport humans and cargo to and beyond low-Earth orbit. Some of

---

2013); GAO, *James Webb Space Telescope: Project Meeting Commitments but Current Technical, Cost, and Schedule Challenges Could Affect Continued Progress*, GAO-14-72 (Washington, D.C.: Jan. 8, 2014); GAO, *NASA: Assessments of Major Projects*, GAO-16-309SP (Washington, D.C.: Mar. 30, 2016); GAO-17-303SP; and GAO-17-317.

[5] GAO-13-4, GAO-13-276SP, GAO-14-72, and GAO-17-303SP.

NASA's projects are expected to incorporate new and sophisticated technologies that must operate in harsh, distant environments.

The life cycle for NASA space flight projects consists of two phases—formulation, which takes a project from concept to preliminary design, and implementation, which includes building, launching, and operating the system, among other activities. NASA further divides formulation and implementation into phase A through phase F. Major projects must get approval from senior NASA officials at key decision points before they can enter each new phase. Figure 1 depicts NASA's life cycle for space flight projects.

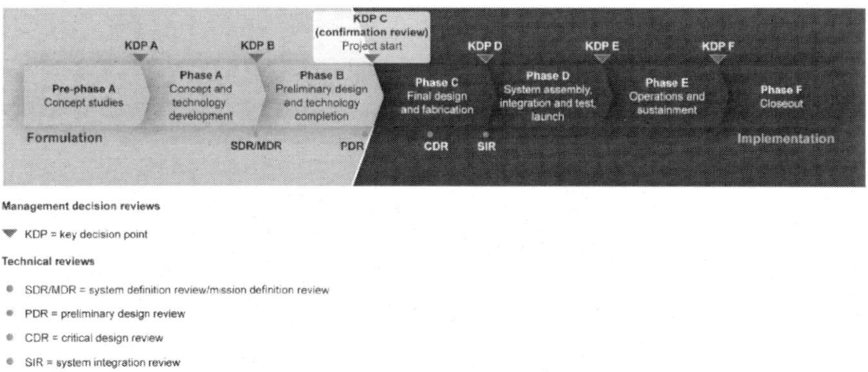

Source: NASA data and GAO analysis. | GAO-18-277I.

Figure 1. NASA's Life Cycle for Space Flight Projects.

Formulation culminates in a review at key decision point C, known as project confirmation, where cost and schedule baselines are established and documented in a decision memorandum. To inform those baselines, each project with a life-cycle cost estimated to be greater than $250 million must also develop a joint cost and schedule confidence level (JCL). The JCL initiative, adopted in January 2009, is a point-in-time estimate that, among other things, includes all cost and schedule elements, incorporates and quantifies known risks, assesses the impacts of cost and schedule to date, and addresses available annual resources. NASA policy requires that projects be baselined and budgeted at the 70 percent confidence level.[6]

The agency baseline commitment established at key decision point C includes cost and schedule reserves held at the project—those within the project manager's control—and NASA headquarters level.[7] Cost reserves are for costs that are expected to be incurred—for instance, to address project risks—but are not yet allocated to a specific part of the project. Schedule reserves are extra time in project schedules that can be allocated to specific activities, elements, and major subsystems to mitigate delays or address unforeseen risks.

STATUS OF NASA'S MAJOR TELESCOPE PROJECTS

NASA's current portfolio of major space telescopes includes three projects—WFIRST, TESS, and JWST—that vary in cost, complexity, and phase of the acquisition life cycle. WFIRST, a project that entered the concept and technology development phase and established preliminary cost and schedule estimates in February 2016, is in the earliest stages of the acquisition life cycle. With preliminary cost estimates ranging from $3.2 billion to $3.8 billion, this project is an observatory designed to perform wide-field imaging and survey of the sky at near-infrared wavelengths to answer questions about the structure and evolution of the universe and to expand our knowledge of planets beyond our solar system. The current design includes a 2.4 meter telescope that was built and qualified for another federal agency over 10 years ago; the project is evaluating which components to reuse and which to modify, refurbish, or build new. TESS—a smaller project whose latest cost estimate is approximately $337 million—is targeted to launch in March 2018 and will be used to conduct the first extensive survey of the sky from space for transiting exoplanets.

---

[6] NASA Procedural Requirements 7120.5E NASA Space Flight Program and Project Management Requirements para 2.4.4 (Aug. 14, 2012) (hereinafter cited as NPR 7120.5E (Aug. 14, 2012). The decision authority for a project can approve it to move forward at less than the 70 percent confidence level. That decision must be justified and documented.

[7] NASA refers to cost reserves as unallocated future expenses.

And finally, JWST, with a life-cycle cost estimate of $8.835 billion, is one of NASA's most complex projects and top priorities. The telescope is designed to help understand the origin and destiny of the universe, the creation and evolution of the first stars and galaxies, and the formation of stars and planetary systems. With a 6.5-meter primary mirror, JWST is expected to operate at about 100 times the sensitivity of the Hubble Space Telescope. JWST's science instruments are to detect very faint infrared sources and, as such, are required to operate at extremely cold temperatures. To help keep these instruments cold, a multi-layered tennis-court-sized sunshield is being developed to protect the mirrors and instruments from the sun's heat.

We have reported for several years on the JWST project, which has experienced significant cost increases and schedule delays. Prior to being approved for development, cost estimates for JWST ranged from $1 billion to $3.5 billion, with expected launch dates ranging from 2007 to 2011. Before 2011, early technical and management challenges, contractor performance issues, low levels of cost reserves, and poorly phased funding levels caused JWST to delay work after confirmation, which contributed to significant cost and schedule overruns, including launch delays. The Chair of the Senate Subcommittee on Commerce, Justice, Science, and Related Agencies requested from NASA an independent review of JWST in June 2010. In response, NASA commissioned the Independent Comprehensive Review Panel, which issued its report in October 2010. The panel concluded that JWST was executing well from a technical standpoint, but that the baseline cost estimate did not reflect the most probable cost with adequate reserves in each year of project execution, resulting in an unexecutable project.[8]

Following this review, Congress in November 2011 placed an $8 billion cap on the formulation and development costs for the project and NASA rebaselined JWST with a life-cycle cost estimate of $8.835 billion that included additional money for operations and a planned launch in

---

[8] James Webb Space Telescope (JWST) Independent Comprehensive Review Panel (ICRP): Final Report (Oct. 29, 2010).

October 2018.[9] The new baseline represented a 78 percent increase to the project's life-cycle cost from the original baseline and a launch date in October 2018, a delay of 52 months. The revised life-cycle cost estimate included a total of 13 months of funded schedule reserve.[10]

Table 2. Current Phase, Cost, and Schedule Status of National Aeronautics and Space Administration's (NASA) Major Space Telescope Projects

Project	Acquisition Phase	Life-Cycle Cost Estimate Preliminary Cost (then-year dollars in millions)			Preliminary Schedule		
Wide-Field Infrared Survey Telescope (WFIRST)	Concept and technology development	3,200-3,800			2024-2026		
		Baseline (then-year dollars in millions)	Latest Estimate (then-year dollars in millions)	Dollar Change (in millions)	Baseline	Target Date	Change (months)
Transiting Exoplanet Survey Satellite (TESS)	System assembly, integration and test, and launch	378.4	336.7	-41.7a	June 2018	March 2018	-3
James Webb Space Telescope (JWST)	System assembly, integration and test, and launch	4,963.6	8,825.4	3,861.8	June 2014	March-June 2019	57-60

Source: GAO analysis of National Aeronautics and Space Administration data. | GAO-18-277T.

[a]NASA decreased TESS's life-cycle cost by $26.7 million after launch vehicle selection in 2014 due to the reduction in planned costs. At its most recent key decision review in August 2017, NASA decreased the project's life-cycle costs again by reallocating $15 million of TESS's headquarters-held reserves to the WFIRST project.

---

[9] A rebaseline is a process initiated if development cost growth is more than 30 percent. This process requires the NASA Administrator to transmit a report to the Committee on Science, Space, and Technology of the House of Representatives and the Committee on Commerce, Science, and Transportation of the U.S. Senate. In addition, if a project or program milestone is likely to be delayed by 6 months or more, a report to the committees is required.

[10] The 2011 rebaseline had 13 months of schedule reserve. However, by accelerating some work, the project was able to increase the schedule reserve to 14 months in June 2012.

Our ongoing work indicates that these three projects are each making progress in line with their phase of the acquisition cycle, but also face challenges in execution. Some of these challenges are unique to the projects themselves and some are common among the projects in NASA's portfolio. For example, when projects enter the integration and test phase, unforeseen challenges can arise and affect the cost and schedule for the project. Table 1 provides more details about the current acquisition phase, cost, and schedule status of NASA's major space telescope projects based on our ongoing work.

WFIRST

NASA's preliminary cost and schedule estimates for the WFIRST project are currently under review as the project responds to findings in the WFIRST Independent External Technical/Management/Cost Review. This independent review was conducted to ensure the mission's scope and required resources are well understood and executable. NASA initiated this review in April 2017 to address the National Academies' concerns that WFIRST cost growth could endanger the balance of NASA's astrophysics program and negatively affect other scientific priorities. The review found that the mission scope is understood, but not aligned with the resources provided and concluded that the mission is not executable without adjustments and/or additional resources. For example, the study team found that NASA's current forecasted funding profile for the WFIRST project would require the project to slow down activities starting in fiscal year 2020, which would result in an increase in development cost and schedule. NASA agreed with the study team's results and directed the project to reduce the cost and complexity of the design in order to maintain costs within the $3.2 billion cost target.

The project is currently identifying potential ways to reduce the scope of planned activities (called "descopes"), assessing the science impact of those descopes, and then developing recommendations for the Astrophysics Division leadership. An example of a descope that may be considered is the requirement for WFIRST to be "star-shade ready," which means the design must be compatible with a star-shade device that is positioned between it and the star being observed to block out starlight while allowing the light emitted by the planet through.

TESS

The TESS project is currently holding cost and schedule reserves consistent with NASA center requirements, but there are no longer headquarters-held cost reserves to cover a delay if the project cannot launch as planned in March 2018.[11] According to a project official, the project is holding 16 days of schedule reserve to its target March 2018 launch readiness date, which includes 6 days for the completion of integration and test, and 10 days for launch operations. The project previously used schedule reserves to accommodate the delayed delivery of its Ka-band transmitter, which is essential for TESS as it transmits the mission data back to Earth, due to continued performance and manufacturing issues. The two main risks to the March 2018 launch date are if: 1) SpaceX requires additional time past December 2017 for NASA's Launch Services Program to certify that TESS can fly on its upgraded launch vehicle—certification is necessary because it will be the first time that NASA will use this version of the vehicle—and 2) any issues are identified during the remainder of environmental testing.

The project is also conducting additional testing on its spare camera at temperatures seen in space to better understand expected camera performance on orbit. TESS will use four identical, wide field-of-view cameras to conduct the first extensive survey of the sky from space for transiting exoplanets. However, during thermal testing, the project found that the substance attaching the lenses to the camera barrel places pressure

---

[11] Both NASA headquarters and the project hold cost reserves for projects. Project-held cost reserves are within the project manager's control. NASA headquarters may allocate cost reserves to the project when project-held cost reserves are not enough to cover additional time and money needed to complete a project or there is an issue outside of the project's control.

on the lenses and causes the cameras to be slightly out of focus. In June 2017, NASA directed the project to proceed with integrating the cameras—as they are expected to meet TESS's top level science requirements even with the anomaly. At its most recent key decision review in August 2017, NASA reallocated $15 million of TESS's headquarters-held reserves to the WFIRST project. While this had the effect of decreasing life cycle costs for TESS, it also increased risk as the project no longer has any additional headquarters-held cost reserves to cover a launch delay past March 2018.

JWST

The JWST project continues to make progress towards launch, but the program is encountering technical challenges that require both time and money to fix and may lead to additional delays, beyond a delay recently announced. While the project has made much progress on hardware integration and testing over the past several months, it also used all of its remaining schedule reserves to address various technical issues, particularly on the spacecraft element. In September 2017, the JWST project requested from the European Space Agency—who will contribute the Ariane V launch vehicle—a launch window from March to June 2019, or 5 to 8 months later than the planned October 2018 launch readiness date, established in 2011. The project based this request on the results of a schedule risk assessment that incorporated inputs from the contractor on expected durations of ongoing spacecraft element integration work and other challenges that were expected to increase schedule.

With the later launch window to June 2019, the project expected to have up to 4 months of new schedule reserves. However, shortly after requesting the revised launch window, the project learned from its contractor that up to another 3 months of schedule reserve use is likely, due to lessons learned from conducting deployment exercises of the sunshield, such as reach and access limitations on the flight hardware. As a result, and pending further examination of the schedule, the project now has approximately one month of schedule reserve to complete environmental testing of the spacecraft element and the final integration phase. The final integration phase is where the instruments and telescope will be integrated with the spacecraft and sunshield to form the completed observatory. As I previously noted, our work has shown the integration and test is the riskiest phase of development, where problems are most likely to be found and schedules slip. Given the risks associated with the integration and test work ahead, coupled with a level of schedule reserves that is currently well below the level stated in the procedural requirements issued by the NASA center responsible for managing JWST, additional delays to the project's revised launch readiness date of June 2019 are likely.[12] As a result, the funding available under the Congressional cost cap of $8 billion may be inadequate as the contractor will need to continue to retain higher workforce levels for longer than expected to prepare the mission for a delayed launch.

LESSONS LEARNED FROM NASA ACQUISITIONS

As Congress, NASA, and the science community consider future telescope efforts, it will be exceedingly important to shape and manage new programs in a manner that minimizes cost overruns and schedule delays. This is particularly important for the largest programs as even small cost increases can have reverberating effects. NASA's telescope and other science projects will always have inherent technical, design, and integration risks because they are complex, specialized, and often push the state of the art in space technology. But too often, our reports find that management and oversight problems—which can include poor planning, optimistic cost estimating, funding gaps, lax oversight, and poor contractor

[12] NASA's Goddard Space Flight Center is the NASA center with responsibility for managing JWST and has issued procedural requirements that establish the levels of both cost and schedule reserves that projects must hold at various points in the project life cycle. Goddard Space Flight Center, Goddard Procedural Requirements 7120.7, Schedule and Budget Margins for Flight Projects (Feb 28, 2017).

performance, among other issues—are the real drivers behind cost and schedule growth.

To its credit, NASA has taken significant steps, partly in response to our past recommendations, to reduce acquisition risk from both a technical and management standpoint, including actions to enhance cost and schedule estimating, provide adequate levels of reserves to projects, establish better processes and metrics to monitor projects, and expand the use of earned value management to better monitor contractor performance. For example, in November 2012, we found that NASA employee skill sets available to analyze and implement earned value management vary widely from center to center, and we recommended that NASA conduct an earned value management skills gap analysis to identify areas requiring augmented capability across the agency, and, based on the results of the assessment, develop a workforce training plan to address any deficiencies.[13] NASA concurred with this recommendation and developed an earned value management training plan in 2014 based on the results of an earned value management skills gap analysis that was conducted in 2013. Moreover, in recent years, we have found that many of the projects within the agency's major project portfolio have improved their cost and schedule performance.[14] Nevertheless, the extent to which NASA has adopted some of the following lessons learned within its portfolio of major projects is mixed, and NASA has an opportunity to strengthen its program management of major acquisitions, including its space telescopes, by doing so.

Manage Cost and Schedule Performance for Large Projects to Limit Implications for Entire Portfolio

In 2013, following JWST's cost increases and schedule growth, we found that though cost and schedule growth can occur on any project,

---

[13] GAO, *NASA: Earned Value Management Implementation across Major Spaceflight Projects Is Uneven*, GAO-13-22 (Washington, D.C.: Nov. 19, 2012).

[14] GAO-16-309SP and GAO-17-303SP.

increases associated with NASA's most costly and complex missions can have cascading effects on the rest of the portfolio.[15] For example, we found that the JWST cost growth would have reverberating effects on the portfolio for years to come and required the agency to identify $1.4 billion in additional resources over fiscal years 2012 through 2017, according to Science Mission Directorate officials. NASA identified approximately half of this required funding from the four science divisions within the Science Mission Directorate account. The majority of the cuts were related to future high priority missions, missions in the operations and sustainment phase, and research and analysis.

In essence, NASA had to mortgage future high priority missions and research to address JWST's additional resource needs. Similarly, the National Academy of Sciences has concluded in the past that it is important for NASA to have a clearly articulated and consistently applied method for prioritizing why and how its scarce fiscal resources are apportioned with respect to the science program in general and on a more granular level among component scientific disciplines. The academy noted that failure to do so could result in a loss of capacity, capability, and human resources in a number of scientific disciplines and technological areas that may take a generation or more to reconstitute once eliminated.[16] NASA's establishment of the WFIRST Independent External Technical/Management/Cost Review that I previously discussed is a step in the right direction to help ensure the Astrophysics Division incorporates this lesson learned.

Establish Adequate Cost and Schedule Reserves to Address Risks

Twice in the history of the JWST program, independent reviewers found that the program's planned cost reserves were inadequate. First, in

---

[15] GAO-13-276SP.

[16] The National Academy of Sciences, *Review of the Draft 2014 Science Mission Directorate Science Plan* (Washington, D.C.: National Academies Press, 2013).

April 2006, an Independent Review Team confirmed that the project's technical content was complete and sound, but expressed concern over the project's reserve funding, reporting that it was too low and phased in too late in the development lifecycle. The review team reported that for a project as complex as JWST, 25 to 30 percent total reserve funding was appropriate. The team cautioned that low reserve funding compromised the project's ability to resolve issues, address risk areas, and accommodate unknown problems. As I previously mentioned, following additional cost increases and schedule threats, NASA commissioned the Independent Comprehensive Review Panel. In 2010, the panel again concluded JWST was executing well from a technical standpoint, but that the baseline cost estimate did not reflect the most probable cost with adequate reserves in each year of project execution, resulting in an unexecutable project.[17]

NASA heeded these lessons when it established a new baseline for JWST in 2011. For example, the revised schedule included more reserves than required by the procedural requirements issued by the NASA center responsible for managing JWST. We have found, however, that NASA has not applied this lesson learned to all of its large projects— most notably with its human spaceflight projects, including the Space Launch System, Orion Crew Capsule, and associated ground systems— and similar outcomes to the JWST project have started to emerge with these projects. We previously reported that all three of these programs were operating with limited cost reserves, which limited each program's ability to address risks and unforeseen technical challenges.

For example, we found in July 2016 that the Orion program planned to maintain very low levels of annual cost reserves until 2018.[18] The lack of available cost reserves in the near term led to the program deferring work to address technical issues to stay within budget, and put the program's future cost reserves at risk of being overwhelmed by deferred work. In April 2017, we also found that all three programs faced development

---

[17] James Webb Space Telescope (JWST) Independent Comprehensive Review Panel (ICRP): Final Report (Oct. 29, 2010).

[18] GAO, *Orion Multi-Purpose Crew Vehicle: Action Needed to Improve Visibility into Cost, Schedule, and Capacity to Resolve Technical Challenges*, GAO-16-620 (Washington, D.C.: July 27, 2016).

challenges in completing work, and each had little to no schedule reserve remaining to the launch date—meaning they would have to complete all remaining work with minimal delay during the most challenging stage of development.[19] We found that it was unlikely that the programs would achieve the planned launch readiness date and recommended that NASA reassess the date. NASA agreed with this recommendation and stated that it would establish a new launch readiness date. In November 2017, NASA announced that a review of the possible manufacturing and production schedule risks indicated a launch date of June 2020—a delay of 19 months—but the agency will manage to a December 2019 launch date because, according to NASA, they have put in mitigation strategies for those risks. We will follow-up on those mitigation strategies as part of future work on the human space exploration programs.

Regularly and Consistently Update Project JCLs to Provide Realistic Estimates to Decision Makers

In 2009, NASA began requiring that programs and projects with estimated life-cycle costs greater than $250 million develop a JCL prior to project confirmation. This was a positive step for NASA to help ensure that cost and schedule estimates are realistic and projects are thoroughly planning for anticipated risks. This is because a JCL assigns a confidence level, or likelihood, of a project meeting its cost and schedule estimates. Our cost estimating best practices recommend that cost estimates should be updated to reflect changes to a program or be kept current as a program moves through milestones.[20] As new risks emerge on a project, an updated cost and schedule risk analysis can provide realistic estimates to decision-makers, including the Congress. This is especially true for NASA's largest

---

[19] GAO, *NASA Human Space Exploration: Delay Likely for First Exploration Mission*, GAO-17-414 (Washington, D.C.: Apr. 27, 2017).

[20] GAO, *GAO Cost Estimating and Assessment Guide: Best Practices for Developing and Managing Capital Program Costs,* GAO-09-3SP (Washington, D.C.: Mar. 2, 2009).

projects as updated estimates may require the Congress to consider a variety of actions.

However, there is no requirement for NASA projects to update their JCLs, and our prior work has found that projects—including JWST—do not regularly update cost risk analyses to take into account newly emerged risks.[21] Our ongoing work indicates that of the 16 major projects currently in NASA's portfolio that have developed JCL estimates, only 2 have reported updating their JCLs (other than required due to a rebaseline). For example, the Interior Exploration using Seismic Investigations, Geodesy, and Heat Transport Project (InSight), a Mars lander, updated its JCL after the project missed its committed launch date. As a result, the project was able to provide additional information to decision makers about the probability that it will meet its revised cost and schedule estimates. As a project reaches the later stages of development, especially integration and testing, the types of risks the project will face may change. An updated project JCL would provide both project and agency management with data on relevant risks that can guide the project decisions. For example, in December 2012, we recommended the JWST project update its JCL.[22] NASA concurred with this recommendation; however, we recently closed the recommendation because NASA had not taken steps to implement it and the amount of time remaining before launch would not have allowed the benefit of implementing the recommendation to be realized. An updated JCL may have portended the current schedule delays, which could have been proactively addressed by the project.

Enhance Oversight of Contractors to Improve Project Outcomes

In December 2012, we found that the JWST project had taken steps to enhance communications with and oversight of its contractors.[23] According

[21] GAO-13-4 and *Space Launch System: Management Tools Should Better Track Cost and Schedule Commitments to Adequately Monitor Increasing Risk,* GAO-15-596 (Washington, D.C.: Jul. 16, 2015).

[22] GAO-13-4.

[23] GAO-13-4.

to project officials, the increased communication allowed them to better identify and manage project risks by having more visibility into contractors' activities. The project reported that a great deal of communication existed across the project prior to the Independent Comprehensive Review Panel; however, additional improvements were made. For example, the project increased its presence at contractor facilities as necessary to provide assistance; this included assigning two engineers on a recurring basis at a Lockheed Martin facility to assist in solving problems with an instrument. The JWST project also assumed full responsibility for the mission system engineering functions from Northrop Grumman in March 2011. NASA and Northrop Grumman officials both said that NASA is better suited to perform these tasks.

We continue to see instances in our ongoing work that highlight the importance of implementing this lesson learned from JWST. For example, we found in 2017 that the Space Network Ground Segment Sustainment project—a project that plans to develop and deliver a new ground system for one Space Network site that provides essential communications tracking services to NASA and non-NASA missions—exceeded its original cost baseline by at least $401.7 million and been delayed by 27 months.[24] The project has attributed some of the cost overruns and schedule delays to the contractor's incomplete understanding of its requirements, which led to poor contractor plans and late design changes. The project also took steps to assign a new NASA project manager, increase physical presence at the contractor facility, and have more staff focused on validation and verification activities.

In summary, NASA continues to make progress developing its space telescopes to help understand the universe and our place in it. But much like other major projects that NASA is developing, there continues to be an opportunity for NASA to learn from JWST and other projects that have suffered from cost overruns and schedule delays. Key project management

[24] GAO-17-303SP. In 2016, NASA announced it was reclassifying SGSS as a hybrid sustainment project for the Space Network. A hybrid sustainment effort is a sustainment effort that still includes development work. The SGSS project expects to experience additional cost growth and schedule delays, but the exact magnitude is unknown. The project was reevaluating its cost and schedules at the time of the review.

tools and prior GAO recommendations that I have highlighted here today, could help to better position these large, complex, and technically challenging efforts for a successful outcome. We look forward to continuing to work with NASA and this subcommittee in addressing these issues.

Chairman Babin, Ranking Member Bera, and Members of the Subcommittee, this completes my prepared statement. I would be pleased to respond to any questions that you may have at this time.

STAFF ACKNOWLEDGMENTS

GAO staff who made key contributions to this statement include Cristina T. Chaplain, Director; Molly Traci, Assistant Director; Richard Cederholm, Assistant Director; Carrie Rogers; Lisa Fisher; Laura Greifner; Erin Kennedy; and Jose Ramos.

In: Commercial Crew Program …
Editor: Graeme N. Lutz
ISBN: 978-1-53614-980-7

Chapter 5

JAMES WEBB SPACE TELESCOPE: INTEGRATION AND TEST CHALLENGES HAVE DELAYED LAUNCH AND THREATEN TO PUSH COSTS OVER CAP*

United States Government Accountability Office

ABBREVIATIONS

ESA	European Space Agency
Goddard	Goddard Space Flight Center
I&T	integration and test
ISIM	Integrated Science Instrument Module
JCL	Joint Cost and Schedule Confidence Level
JWST	James Webb Space Telescope
NASA	National Aeronautics and Space Administration

* This is an edited, reformatted and augmented accessible version of the United States Government Accountability Office Report to Congressional Committees, Publication No. GAO-18-273, dated February 2018.

OTE	Optical Telescope Element
OTIS	Optical Telescope Element and Integrated Science Instrument Module
STScI	Space Telescope Science Institute

Why GAO Did This Study

JWST, a large, deployable telescope intended to be the successor to the Hubble Space Telescope, is one of NASA's most complex and expensive projects, at an anticipated cost of $8.8 billion. Congress set an $8 billion JWST development cost cap in 2011, and the remaining $837 million is for its operations costs. JWST is intended to revolutionize our understanding of star and planet formation and advance the search for the origins of our universe. With significant integration and testing planned for the remaining period until launch, the JWST project will still need to address many challenges during the remainder of integration and testing.

Conference Report No. 112-284, accompanying the Consolidated and Further Continuing Appropriations Act, 2012, included a provision for GAO to assess the project annually and report on its progress. This is the sixth such report. This chapter assesses the extent to which JWST is (1) meeting its schedule commitments, and (2) able to meet its cost commitments. GAO reviewed monthly JWST reports, reviewed relevant policies, conducted independent analysis of NASA and contractor data, and interviewed NASA and contractor officials.

What GAO Recommends

GAO has made recommendations on the project in previous reports. NASA agreed with and took action on many of GAO's prior recommendations, but not on others—some of which may have provided insight to the current schedule delays. For example, in December 2012,

GAO recommended that the JWST project perform an updated integrated cost/schedule risk analysis.

WHAT GAO FOUND

In 2017, the National Aeronautics and Space Administration's (NASA) James Webb Space Telescope (JWST) project delayed its launch readiness date by at least 5 months, and further delays are likely. The delay—from October 2018 to a launch window between March and June 2019—was primarily caused by components of JWST's spacecraft taking longer to integrate than planned. JWST made considerable progress toward the completion of integration and test activities in the past year. However, the project used all remaining schedule reserve—or extra time set aside in the schedule in the event of delays or unforeseen risks—to address technical issues, including an anomaly on the telescope found during vibration testing. Extending the launch window provided the project up to 4 months of schedule reserve. However, shortly after requesting the new launch window in September 2017, the project determined that several months of schedule reserve would be needed to address lessons learned from the initial folding and deployment of the observatory's sunshield (see image). Given remaining integration and test work ahead—the phase in development where problems are most likely to be found and schedules tend to slip—coupled with only 1.5 months of schedule reserves remaining to the end of the launch window, additional launch delays are likely. The project's Standing Review Board will conduct an independent review of JWST's schedule status in early 2018 to determine if the June 2019 launch window can be met.

JWST will also have limited cost reserves to address future challenges, such as further launch delays, and is at risk of breaching its $8 billion cost cap for formulation and development set by Congress in 2011. For several years, the prime contractor has overestimated workforce reductions, and technical challenges have prevented these planned reductions, necessitating the use of cost reserves. Program officials said that existing program resources will accommodate the new launch window— provided remaining integration and testing proceeds as planned without any long delays. However, JWST is still resolving technical challenges and work continues to take longer than planned to complete. As a result, the project is at risk of exceeding its $8 billion formulation and development cost cap.

Source: Northrop Grumman. | GAO-18-273.

James Webb Space Telescope's Sunshield Folding Operation at the Contractor Facility

February 28, 2018
Congressional Committees

The James Webb Space Telescope (JWST) is one of the National Aeronautics and Space Administration's (NASA) most complex projects and top priorities, which is intended to revolutionize our understanding of star and planet formation and advance the search for the origins of our universe. The innovative technologies within the telescope, as well as the sheer size of some of its components—such as the tennis-court-sized sunshield—illustrate some of the immense development challenges. The project is now in the midst of a period of significant integration and test activities for its elements and major subsystems. During this period, unforeseen challenges could arise and affect the cost and schedule for the project, as evidenced by a recent announcement that JWST delayed its launch readiness date by 5 to 8 months because of technical challenges. Until its launch, NASA and its contractors' continued ability to identify and respond to challenges in a timely and cost-effective manner will likely influence whether JWST can meet its cost commitments to Congress and avoid further schedule delays.

The on-time and on-budget delivery of JWST is also a high congressional priority, as Conference Report No. 112-284 included a provision for GAO to assess the JWST program annually and to report to the Committees on Appropriations on key issues relating to program and risk management, achievement of cost and schedule goals, program technical status, and oversight mechanisms.[1] This chapter is our sixth in response to that provision. For this chapter, we assessed the extent to which (1) the JWST project is managing technical challenges and integration risks in order to meet its schedule commitments, and (2) technical challenges and higher than planned contractor workforce levels are affecting the JWST project's ability to meet its cost commitments.

To address these objectives, we examined the schedule, technical, and cost performance of the project since our last report in December 2016—which also focused on the project's cost and schedule commitments.[2] To assess the extent to which the JWST project is managing technical challenges and integration risks in order to meet its schedule commitments, we reviewed project and contractor schedule documentation, and held interviews with program, project, and contractor officials on the progress made and any challenges faced building and integrating the different components of the observatory. We examined and analyzed monthly project status reports to management to monitor schedule reserve levels and usage and potential risks and technical challenges that may impact the project's schedule, and to gain insights on the project's progress since our last report in December 2016. Further, we attended flight program reviews at the National Aeronautics and Space Administration (NASA)

[1] H.R. Conf. Rep. No. 112-284, at 254 (2011).

[2] GAO, *James Webb Space Telescope: Project Meeting Cost and Schedule Commitments but Continues to Use Reserves to Address Challenges*, GAO-17-71 (Washington, D.C.: Dec. 7, 2016).

headquarters on a quarterly basis, where the current status of the program was briefed to NASA headquarters officials outside of the project. We examined selected individual risks for elements and major subsystems from monthly risk registers prepared by the project to understand the likelihood of occurrence and impacts to the schedule based on steps the project is taking to mitigate the risks. We examined the project's schedule risk assessment to determine the impact of technical challenges and schedule execution rates on the project's planned launch readiness date. Furthermore, we interviewed project officials at Goddard Space Flight Center, contractor officials from the Northrop Grumman Corporation, and the Association of Universities for Research in Astronomy's Space Telescope Science Institute concerning technological challenges that have had an impact on schedule, and the project's and contractor's plans to address these challenges. We also obtained information from independent NASA reviewers on the status of project software and challenges.

To assess the extent to which technical challenges and higher than planned contractor workforce levels are affecting the JWST project's ability to meet its cost commitments, we reviewed and analyzed program, project, and contractor data and documentation and held interviews with officials from these organizations. We reviewed JWST project status reports on cost issues to determine the risks that could impact cost. We analyzed contractor workforce plans against workforce actuals to determine whether contractors are meeting their workforce plans. We monitored and analyzed the status of program and project cost reserves in current and future fiscal years to determine the project's financial posture. We examined and analyzed earned value management data from the project's observatory contractor to identify trends in performance, whether tasks were completed as planned, and likely estimates at completion.

Our work was performed primarily at NASA headquarters in Washington, D.C.; Goddard Space Flight Center in Greenbelt, Maryland; Northrop Grumman Corporation in Redondo Beach, California; and the Space Telescope Science Institute in Baltimore, Maryland. We also obtained information from officials at the Independent Verification and Validation facility in Fairmont, West Virginia.

We conducted this performance audit from March 2017 to February 2018 in accordance with generally accepted government auditing standards. Those standards require that we plan and perform the audit to obtain sufficient, appropriate evidence to provide a reasonable basis for our findings and conclusions based on our audit objectives. We believe that the evidence obtained provides a reasonable basis for our findings and conclusions based on our audit objectives.

BACKGROUND

JWST is envisioned to be a large deployable space telescope, optimized for infrared observations, and the scientific successor to the aging Hubble Space Telescope. JWST is being designed for a 5-year mission to find the first stars, study planets in other solar systems to search for the building blocks of life elsewhere in the universe, and trace the evolution of galaxies from their beginning to their current formation. JWST is intended to operate in an orbit approximately 1.5 million kilometers—or 1 million miles—from the Earth. With a 6.5-meter primary mirror, JWST is expected to operate at about 100 times the sensitivity of the Hubble Space Telescope. JWST's science instruments are designed to observe very faint infrared sources and therefore are required to operate at extremely cold temperatures. To help keep these instruments cold, a multi-layered tennis court-sized sunshield is being developed to protect the mirrors and instruments from the sun's heat.

The JWST project is divided into three major segments: the observatory segment, the ground segment, and the launch segment. When complete, the observatory segment of JWST is to include several elements (Optical Telescope Element (OTE), Integrated Science Instrument Module (ISIM), and spacecraft) and major subsystems (sunshield and cryocooler).[3] The hardware configuration created when the Optical Telescope Element and

[3] The cryocooler is an interdependent two-stage cooler subsystem designed to bring the infrared light detector within JWST's Mid-Infrared Instrument to the required temperature of 6.7 Kelvin.

the Integrated Science Instrument Module were integrated, referred to as OTIS, is not considered an element by NASA, but we categorize it as such for ease of discussion. Additionally, JWST is dependent on software to deploy and control various components of the telescope, and to collect and transmit data back to Earth. The elements, major subsystems, and software are being developed through a mixture of NASA, contractor, and international partner efforts. See figure 1 for an interactive graphic that depicts the elements and major subsystems of JWST.

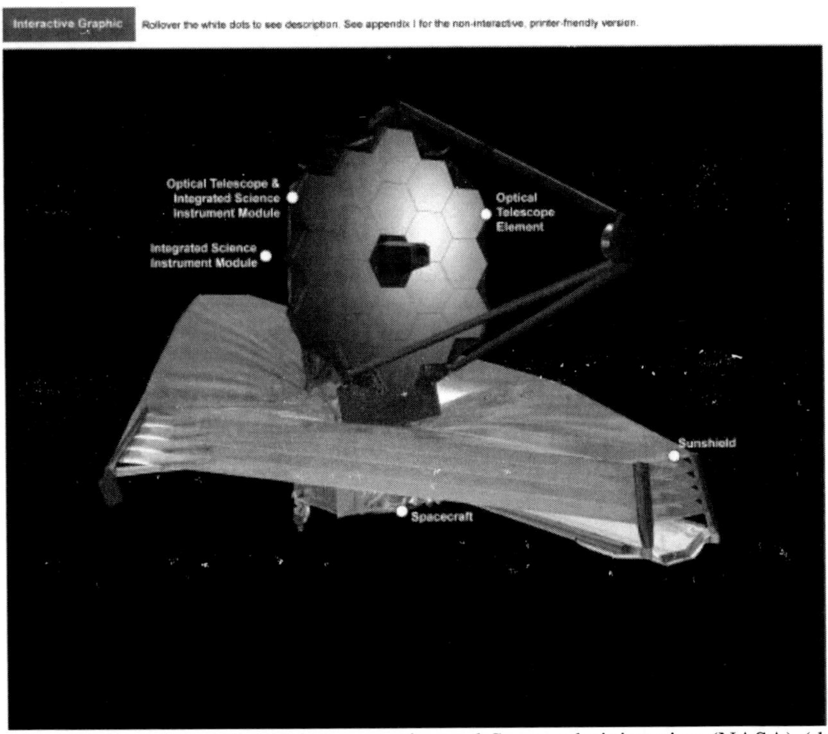

Source: GAO analysis; National Aeronautics and Space administration (NASA) (data and images). | GAO-18-273.

Figure 1. James Webb Space Telescope.

For the majority of work remaining, the JWST project is relying on two contractors: Northrop Grumman Corporation and the Association of Universities for Research in Astronomy's Space Telescope Science

Institute (STScI). Northrop Grumman plays the largest role, developing the sunshield, the Optical Telescope Element, the spacecraft, and the Mid-Infrared Instrument's cryocooler, in addition to integrating and testing the observatory. STScI's role includes soliciting and evaluating research proposals from the scientific community, and receiving and storing the scientific data collected, both of which are services that it currently provides for the Hubble Space Telescope. Additionally, STScI is developing the ground system that manages and controls the telescope's observations and will operate the observatory on behalf of NASA. JWST will be launched on an Ariane 5 rocket, provided by the European Space Agency.

JWST depends on 22 deployment events—more than a typical science mission—to prepare the observatory for normal operations on orbit. For example, the sunshield and primary mirror are designed to fold and stow for launch and deploy once in space. Due to its large size, it is nearly impossible to perform deployment tests of the fully assembled observatory, so the verification of deployment elements is accomplished by a combination of lower level component tests in flight-simulated environments; ambient deployment tests for assembly, element, and observatory levels; and detailed analysis and simulations at various levels of assembly.

Schedule and Cost Reserves for NASA Projects

We have previously reported that complex development efforts like JWST face numerous risks and unforeseen technical challenges, which oftentimes can become apparent during integration and testing. To accommodate unanticipated challenges and manage risk, projects reserve extra time in their schedules, which is referred to as schedule reserve, and extra money in their budgets, which is referred to as cost reserve.

Schedule reserve is allocated to specific activities, elements, and major subsystems in the event of delays or to address unforeseen risks. Each

JWST element and major subsystem has been allocated schedule reserve. When an element or major subsystem exhausts schedule reserve, it may begin to affect schedule reserve on other elements or major subsystems whose progress is dependent on prior work being finished for its activities to proceed. The element or major subsystem with the least amount of schedule reserve determines the critical path for the project. Any delay to an activity that is on the critical path will reduce schedule reserve for the whole project, and could ultimately impact the overall project schedule.

Cost reserves are additional funds within the project manager's budget that can be used to address unanticipated issues for any element or major subsystem, and are used to mitigate issues during the development of a project. For example, cost reserves can be used to buy additional materials to replace a component or, if a project needs to preserve schedule reserve, reserves can be used to accelerate work by adding shifts to expedite manufacturing. NASA's Goddard Space Flight Center (Goddard)—the NASA center with responsibility for managing JWST— has issued procedural requirements that establish the levels of both cost and schedule reserves that projects must hold at various phases of development.[4] In addition to cost reserves held by the project manager, management reserves are funds held by the contractors that allow them to address cost increases throughout development. We have found that management reserves should contain 10 percent or more of the cost to complete a project and are generally used to address various issues tied to the contract's scope.[5]

---

[4] NASA, Goddard Procedural Requirements 7120.7, Schedule and Budget Margins for Flight Projects (May 4, 2008); Goddard projects are required to hold two months of funded schedule reserve per year from the start of integration and test to shipment to launch site, and are required to hold cost reserves equal to at least 25 percent at the project confirmation review, and 10 percent or higher at the time of delivery to the launch site.

[5] GAO, NASA: Earned Value Management Implementation across Major Spaceflight Projects Is Uneven, GAO-13-22 (Washington, D.C.: Nov. 19, 2012); and GAO Cost Estimating and Assessment Guide: Best Practices for Developing and Managing Capital Program Costs, GAO-09-3SP (Washington, D.C.: Mar. 2, 2009).

History of Cost Growth and Schedule Delays

JWST has experienced significant cost increases and schedule delays. Prior to being approved for development, cost estimates of the project ranged from $1 billion to $3.5 billion, with expected launch dates ranging from 2007 to 2011. Before 2011, early technical and management challenges, contractor performance issues, low level cost reserves, and poorly phased funding levels caused JWST to delay work after cost and schedule baselines were established, which contributed to significant cost and schedule overruns, including launch delays.

The Chair of the Senate Subcommittee on Commerce, Justice, Science, and Related Agencies requested from NASA an independent review of JWST in June 2010. NASA commissioned the Independent Comprehensive Review Panel, which issued its report in October 2010, and concluded that the baseline funding did not allot adequate reserves, resulting in an unexecutable project.[6] Following this review, the JWST program underwent a replan in September 2011, and in November of that same year, Congress placed an $8 billion cap on the formulation and development costs for the project. On the basis of the replan, NASA rebaselined JWST with a life-cycle cost estimate of $8.835 billion which included additional money for operations and a planned launch in October 2018.[7] The revised life-cycle cost estimate included a total of 13 months of funded schedule reserve.[8]

---

[6] James Webb Space Telescope (JWST) Independent Comprehensive Review Panel (ICRP): Final Report (Oct. 29, 2010).

[7] A replan is a process initiated if development cost growth is 15 percent or more. A replan does not require a new project baseline to be established. A rebaseline is a process initiated if development cost growth is more than 30 percent. Both processes require the NASA Administrator to transmit a report to the Committee on Science, Space, and Technology of the House of Representatives and the Committee on Commerce, Science, and Transportation of the Senate. In addition, if a project or program milestone is likely to be delayed by 6 months or more, a report to the committees is required.

[8] The 2011 baseline plan had 13 months of schedule reserve. However, by accelerating some work, the project was able to increase the schedule reserve to 14 months in June 2012.

Previous GAO Reviews of JWST Project

We have previously found that since the project's replan in 2011, the JWST project has met its cost and schedule commitments.[9] In our most recent report in December 2016, we found that the project was still operating within its committed schedule while in its riskiest phase of development—integration and test—but had used about 3 months of schedule reserve since our previous December 2015 report.[10] In addition, we found that the project was facing numerous risks and single points of failure before launch. Finally, we found that while the project was meeting its cost commitments despite technical and workforce challenges, the observatory contractor had continued to maintain a larger workforce for longer than planned in order to address technical issues. In these prior reports, we have made recommendations with regard to improving cost and schedule estimating, updating risk assessments, and strengthening management oversight. NASA has generally agreed and taken steps to implement a number of our recommendations. For example, in December 2015, we recommended that the JWST project require contractors to identify, explain, and document anomalies in contractor-delivered monthly earned value management reports. NASA concurred with this recommendation and, in February 2016, directed the contractors to implement the actions stated in the recommendation. However, NASA did not implement some recommendations, which if implemented, may have provided insight into the challenges it now faces. For example, in December 2012, we recommended the JWST project update its joint cost and schedule confidence level (JCL), a point-in-time estimate that, among

---

[9] GAO, *James Webb Space Telescope: Project Meeting Cost and Schedule Commitments but Continues to Use Reserves to Address Challenges*, GAO-17-71 *(Washington, D.C.: Dec. 7, 2016); James Webb Space Telescope: Project on Track but May Benefit from Improved Contractor Data to Better Understand Costs*, GAO-16-112 (Washington, D.C.: Dec. 17, 2015); *James Webb Space Telescope: Project Facing Increased Schedule Risk with Significant Work Remaining*, GAO-15-100 (Washington, D.C.: Dec. 15, 2014); *James Webb Space Telescope: Project Meeting Commitments but Current Technical, Cost, and Schedule Challenges Could Affect Continued Progress*, GAO-14-72 (Washington, D.C.: Jan. 8, 2014); and *James Webb Space Telescope: Actions Needed to Improve Cost Estimate and Oversight of Test and Integration*, GAO-13-4 (Washington, D.C.: Dec. 3, 2012).

[10] GAO-17-71.

other things, includes all cost and schedule elements and incorporates and quantifies known risks.[11] NASA policy requires projects to establish commitment baselines at a 70 percent confidence level. Although NASA concurred with this recommendation, it did not take steps to implement it. An updated JCL may have portended the current schedule delays, which could have been proactively addressed by the project.

CONSIDERABLE PROGRESS MADE ACROSS JWST PROJECT, BUT INTEGRATION AND TEST CHALLENGES HAVE DELAYED LAUNCH AT LEAST 5 MONTHS WITH FURTHER DELAYS LIKELY

While much progress on hardware integration and testing and several risk reduction efforts have occurred over the past several months, the JWST project also used all of its schedule reserves established at the replan in 2011 to address various technical issues, including a test anomaly on the telescope and sunshield hardware challenges. In September 2017, the JWST project requested a launch window at least 5 to 8 months later than the planned October 2018 launch readiness date, based on the results of a schedule risk assessment that showed that various components of the spacecraft element integration were taking longer to complete than expected. The new launch window included up to 4 months of additional schedule reserves. However, shortly after requesting the revised launch window from the European Space Agency (ESA), which will contribute the launch vehicle, the project learned from Northrop Grumman that up to another 3 months of schedule reserve use was expected, due to lessons learned from conducting deployment exercises of the spacecraft element and sunshield. After incorporating some schedule efficiencies, the project now has 1.5 months of schedule reserve remaining. Given the remaining integration and test work ahead—the phase in development where

---

[11] GAO-13-4.

problems are most likely to be found and schedules tend to slip—and risks remaining to be reduced to acceptable levels, coupled with a low level of schedule reserves, we believe that additional delays to the project's launch readiness date are likely.

JWST Project Completed Significant Integration and Test Work

Since our last report, the JWST project has made considerable progress toward completing its third and fourth of five total integration and test phases for the combined optical telescope element and integrated science instrument module (OTIS) and the spacecraft elements, respectively. Previously, the project and Northrop Grumman completed the Integrated Science Instrument Module and the Optical Telescope Element integration phases in March 2016, as shown in Figure 2 below.

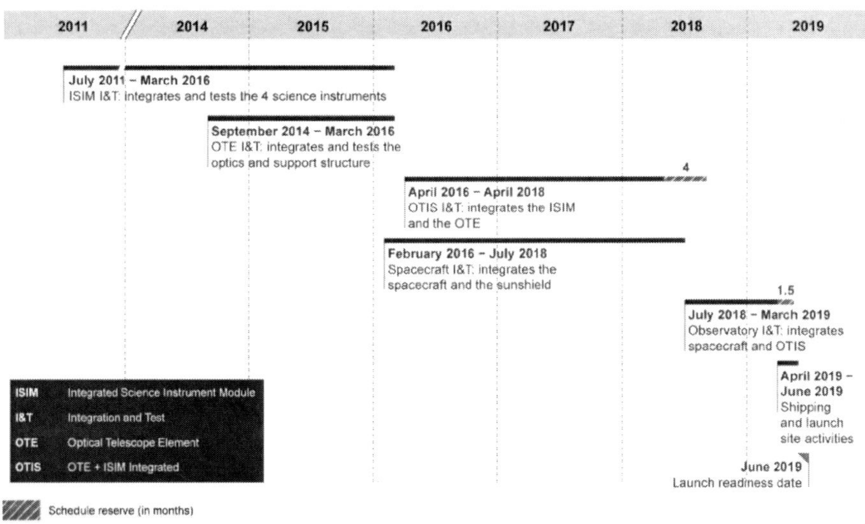

Source: GAO analysis; National Aeronautics and Space administration (NASA) data. | GAO-18-273.

Figure 2. Integration and Test (I&T) Schedule for the James Webb Space Telescope.

- *OTIS progress:* Hardware integration and two of three key environmental tests—acoustics and vibration—were completed in 2016 and early 2017, respectively. The third key test, cryovacuum— which was conducted in a large cryovacuum chamber to ensure the telescope can operate at the near-absolute zero cryogenic temperatures of space—began in July 2017 at Johnson Space Center and successfully concluded in October 2017. The project identified a technical issue with the stability of the optical mirror that affects image quality, and by conducting some additional testing, determined that it was caused by a test equipment setup issue and not related to the flight hardware itself. Project officials stated that they plan to delay shipping the completed OTIS element to the Northrop Grumman facility in California for final integration with the spacecraft element from late December 2017 to February 2018. According to project officials, the delay allows the project to shift some of the work to prepare OTIS for integration with the spacecraft—such as cleaning the mirrors—to Johnson Space Center where it will not have to share space in the crowded clean room at Northrop Grumman as sunshield fold and stow activities are ongoing. OTIS is expected to arrive at Northrop Grumman months ahead of its need date for integration into the observatory.
- *Spacecraft element progress:* All spacecraft element hardware has been delivered and mechanical integration of spacecraft hardware— including the five layers of the sunshield—is largely complete. Northrop Grumman has also completed a folding operation and the first full deployment of the integrated spacecraft element. Northrop Grumman plans to refold the sunshield and complete one more deployment cycle, after environmental testing, in this phase of integration and testing.

The project and its contractors conducted risk reduction testing on OTIS and the spacecraft elements to reduce risk for challenging environmental tests on flight hardware.

Source: National Aeronautics and Space Administration. | GAO-18-273.

Figure 3. Optical Telescope Element and Integrated Science Instrument Module (OTIS) Emerging from Thermal Vacuum Testing Chamber.

Source: Northrop Grumman. | GAO-18-273.

Figure 4. Flight Sunshield Deployed and Tensioned at Northrop Grumman Facility.

These tests allowed the project and its contractors to practice processes and procedures for testing on flight hardware to create a more efficient test flow and proactively address issues before flight hardware tests commenced. For example, the second risk reduction test on the OTIS pathfinder hardware showed that vibration levels inside the test chamber were too high, and adjustments to the ground support equipment were implemented to address this issue. Additionally, Northrop Grumman officials noted that risk reduction tests on the spacecraft element have helped demonstrate facility capability and logistics for the upcoming tests of flight hardware.

The project has also progressed in preparing the software and ground systems that will operate the observatory and manage and control the telescope's observations. According to NASA's Independent Verification and Validation group, the overall status of JWST software development and integration efforts is very positive with minimal development remaining, and the group has significant confidence that the mission software will support the mission objectives. Additionally, the Space Telescope Science Institute has made considerable progress in preparing JWST's ground systems, such as preparing the Mission Operations Center and conducting the Mission Operations Review in April 2017.

The project has made notable progress in reducing and closing numerous tracked risks. In December 2016, we reported that the project maintained a risk list with 73 items. Currently, the list of tracked risks has 47 items to be closed or mitigated to acceptable levels. The completion of the OTIS cryovacuum test enabled the project to recently close several risks. For example, the project previously tracked a risk that the instrument module and telescope element might have to be de-integrated if OTIS testing revealed workmanship issues. With the successful completion of the testing, this risk was closed in fall 2017. The project also obtained a waiver from the Office of Safety and Mission Assurance to NASA's risk policy for its over 300 single point failures throughout the observatory, the majority of which are related to the sunshield. Project officials reported that the elimination of all single point failures on the JWST Mission is not practical or even feasible, due mainly to the large number of deployments, and that all mitigations practical to address and minimize them have been implemented.

JWST Delayed Launch Due to Integration Challenges on the Spacecraft Element, Avoiding a Potential Launch Site Conflict

In the summer of 2017, the JWST project conducted a schedule risk assessment that showed that the October 2018 launch readiness date was unachievable, primarily due to the various components of spacecraft element integration taking longer to complete than planned. The project performed the schedule risk assessment in order to provide ESA a desired launch window about one year prior to the expected launch date. The assessment took into account remaining work to be completed, lessons learned from environmental testing, and the current performance rates of integrating the spacecraft element. As a result of the assessment, in September 2017 NASA requested from ESA a launch window of March 2019 to June 2019. The requested launch window represents a 5- to 8-month delay from the previously planned October 2018 launch readiness date.

The schedule risk assessment incorporated input from Northrop Grumman on expected durations for remaining spacecraft and observatory level integration activities. However, the project's analysis determined that the expected durations provided by Northrop Grumman were overly optimistic. As a result, the project incorporated uncertainty factors into the analysis, which added 2 to 3 months to the schedule. The project also estimated an additional 5 to 8 weeks would be needed because of emerging technical issues not specifically accounted for by the schedule risk assessment. Additionally, the project updated the expected time required at the launch site for processing activities and added about 1.25 months. According to project officials, the confidence in the launch window identified is in line with that of a typical NASA JCL at 70 percent. NASA's independent Standing Review Board reviewed the assessment and found that it was a thorough approach for reviewing the schedule, risks, and uncertainties and that the new proposed launch readiness range is technically feasible with reasonable risk.

NASA's request for a March to June 2019 launch window was driven by its own schedule and technical issues, but also avoids potential conflicts with other mission launches. Regardless of JWST's launch readiness, and prior to undertaking the schedule risk assessment, the project learned in November 2016 of potential scheduling conflicts at the launch site in French Guiana. After numerous delays, BepiColombo, a joint ESA/Japan Aerospace Exploration Agency mission to Mercury, is currently forecasted to have an October 2018 launch readiness date. According to program officials, that mission could have taken precedence over JWST given that planetary missions generally have more limited launch windows. Additionally, Arianespace, a commercial company, currently has a commercial launch scheduled for the December 2018 timeframe.

JWST Project Consumed all of its Planned Schedule Reserve to Address Technical Challenges

While much progress has been made since we last reported in December 2016, the project and Northrop Grumman consumed the remaining 6 months of schedule reserves established at the 2011 replan to address technical challenges that arose during the OTIS and the spacecraft element integration and test work, as well as additional challenges identified by the schedule risk assessment. Specifically:

- In February 2017, a vibration anomaly during OTIS vibration testing at Goddard Space Flight Center, occurring in parallel with spacecraft and sunshield issues, consumed 1.25 months and delayed the start of cryovacuum testing, the final event in the OTIS integration and test phase, by several weeks.
- In April 2017, spacecraft and sunshield issues consumed an additional 1.25 months. Specifically, a contractor technician applied too much voltage and irreparably damaged the spacecraft's pressure transducers, components of the propulsion system, which

help monitor spacecraft fuel levels. The transducers had to be replaced and reattached in a complicated welding process. At the same time, Northrop Grumman also addressed several challenges with integrating sunshield hardware such as the mid-boom assemblies and membrane tensioning system, which help deploy the sunshield and maintain its correct shape.

- Finally, in September 2017, the remaining 3.5 months of previously planned schedule reserves was consumed as a result of the contractor having underestimated the time required to complete integration and test work on the spacecraft and other risks identified in the schedule risk analysis. Specifically, execution of spacecraft integration and test tasks, due to the complexity of work and cautious handling given sensitivity of flight hardware, was slower than planned. For example, the installation of numerous membrane retention devices slowed the pace of the work. According to Northrop Grumman officials, the sunshield is elevated off the ground for installation work and the size and quantity of the work lifts necessary for the technicians to access the sunshield requires more maneuvering and prevents the technicians from working on the forward and aft sunshield assemblies simultaneously.

Taking into account the consumption of planned reserves and the establishment of the revised launch window, the project expected to have up to 4 months of schedule reserve extending to the end of the launch window range, or June 2019. However, shortly after the project notified ESA of the launch delay in September 2017, the project received updated information from Northrop Grumman and determined that up to 3 months of schedule reserve would be needed based upon lessons learned from Northrop Grumman's initial sunshield folding operation and implications for remaining deployment test activities. After incorporating some schedule efficiencies, the project now has 1.5 months of schedule reserve remaining. This level of schedule reserve is below the standards established by Goddard Space Flight Center for a project at this stage of development. The project is working with Northrop Grumman to determine if any further schedule reserve can be regained by incorporating schedule efficiencies and adjusting integration and test plans. As shown in the figure below, Northrop Grumman's work on the spacecraft element remains on the project's critical path—the schedule with the least amount of reserve, which determines the overall schedule reserve for the project—now with an estimated 1.5 months of schedule reserve to the end of the launch window in June 2019.

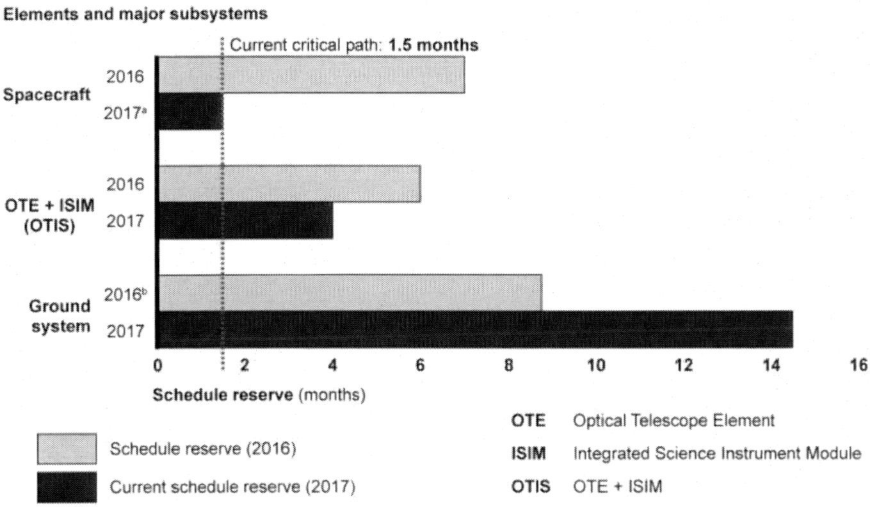

Source: GAO analysis of National Aeronautics and Space Administration (NASA) data. | GAO-18-273.

[a] In 2017, the spacecraft was integrated with the sunshield to form the spacecraft element.

[b] Ground system schedule as reported in February 2017.

Figure 5. Schedule Reserve Held by Elements and Major Subsystems for the James Webb Space Telescope in 2016 and 201.

Ongoing Spacecraft Integration and Test Issues, Challenging Remaining Work, and Slow Contractor Performance Make Additional Launch Delays Likely

Given several ongoing technical issues, and the work remaining to test the spacecraft element and complete integration of the telescope and spacecraft, combined with continuing slower than planned work at Northrop Grumman, we believe that the rescheduled launch window is likely unachievable. For example, in May 2017, Northrop Grumman found that 8 of 16 valves in the spacecraft propulsion system's thruster modules were leaking beyond allowable levels. The project and Northrop Grumman were unable to definitively isolate the root cause of the leaks; however, Northrop Grumman determined that the most likely cause is a handling error at their facility. Specifically, the material around the valves deteriorated due to a solvent used for cleaning. All of the thruster modules were returned to the vendor for investigation and refurbishment.

According to project officials, the refurbished thruster modules were returned to the contractor facility in late 2017 for reattachment. However, reattaching the repaired modules is a challenge because of the close proximity of electronics and other concerns. The project included about one month in the schedule risk assessment to account for the time spent investigating and determining the path forward for the thruster issue; however, the full schedule impact of reattaching the thruster modules to the spacecraft element had not yet been determined and was not incorporated into the analysis. In November 2017, the project and Northrop Grumman chose a reattachment method that project officials stated is expected to require less time to complete and pose fewer risks to the hardware than a traditional welding approach.

In October 2017, when conducting folding and deployment exercises on the sunshield, Northrop Grumman discovered several tears in the sunshield membrane layers. According to program officials, a workmanship error contributed to the tears. The tears can be repaired; however, some schedule reserve may be required to repair them. Additionally, during the deployment exercise, one of the sunshield's six membrane tensioning systems experienced a snag. Northrop Grumman is planning to implement a slight design modification to prevent the issue from occurring again. Northrop Grumman officials have not yet determined if the schedule will be affected as a result.

Beyond mitigating the specific spacecraft thruster module valve leak and sunshield issues, the project faces significant work ahead, and numerous risks remain to be mitigated to acceptable levels. For example, the project and Northrop Grumman must:

- Resolve lingering technical issues from the OTIS cryovacuum test and prepare and ship OTIS to the Northrop Grumman facility in California for integration with the spacecraft.
- Complete integration of spacecraft hardware, and conduct spacecraft element environmental tests and remaining deployments of the spacecraft and sunshield—activities which, to date, have taken considerably longer than planned.
- Integrate the completed OTIS element with the spacecraft element and test the full observatory in the fifth and final integration phase, which includes another set of challenging environmental tests.
- Mitigate approximately 47 remaining tracked hardware and software risks to acceptable levels and continue to address the project's 300+ potential single point failures to the extent possible.
- Prepare and ship the observatory to the launch site and complete final launch site processing, including installation of critical release mechanisms.

Project officials have expressed concern with Northrop Grumman's ability to prevent further schedule erosion as the project moves through remaining integration and test work. With the project's current low level of schedule reserves, even a relatively minor disruption could cause the project to miss its revised launch window. According to program officials, the contractor has increased its daily work shifts from two to three and is now working 24 hours per day on spacecraft integration, which further limits schedule flexibility. In early 2018, the project's independent Standing Review Board will review the latest schedule inputs based on updated knowledge about spacecraft integration and test activity durations. For example, according to project officials, by early 2018, the contractor is expected to have completed the second of four planned fold and stow sequences on the sunshield, which will provide more insight into whether the current planned schedule is realistic. The Standing Review Board will also examine the project's plans for schedule efficiencies and potential integration and test adjustments to determine if the June 2019 launch window can be met. Project officials stated that following this review, NASA senior management will be briefed on the Standing Review Board's findings and will then formally identify a new launch readiness window.

Our prior work has shown that integration and testing is the phase in which problems are most likely to be found and schedules tend to slip. For a uniquely complex project such as JWST, this risk is magnified. Now that the project is well into its complex integration and test efforts, events are sequential in nature and there are fewer opportunities to mitigate issues in parallel. Since the replan, the project has used about 2.5 months of schedule reserve per year to address technical issues, but, as discussed above, it now has only approximately 1.5 months of schedule reserve to last until the end of the revised launch window in June 2019. Thus, past experience with technical issues in earlier integration phases suggests that this amount of reserve will not be adequate for the challenging work ahead, and further delays to launch readiness are likely. We will continue to monitor the project's progress in meeting its revised schedule as more information is available during this critical integration and test phase.

Higher Contractor Workforce Levels to Address Continuing Technical Challenges Places JWST at Risk of Exceeding Cost Commitments

Northrop Grumman continued to maintain higher than planned workforce levels in the past year and, as a result, NASA will have limited cost reserves to address future challenges. Northrop Grumman's ability to control costs and decrease its workforce is central to JWST's capacity to meet its long-term cost commitments. For the past 44 months, Northrop Grumman's actual workforce has exceeded its projections and the company is not expected to significantly reduce its workforce until the spring of 2019, when NASA plans to ship the completed observatory to the launch site. Northrop Grumman had planned to reduce its workforce in fiscal years 2016 and 2017 as work was planned to be completed, but has needed to maintain higher workforce levels due to technical challenges and the work taking longer than expected. Figure 6 illustrates the difference between the workforce levels that Northrop Grumman projected for fiscal years 2016 and 2017, and its actual workforce levels during that period.

As shown in figure 6, Northrop Grumman has slightly reduced its workforce since the beginning of fiscal year 2016. However, staffing levels remain higher than projected as a result of previously noted technical challenges including spacecraft and sunshield integration and test challenges, to keep specialized engineers available when needed during final assembly, and to complete required testing activities. Projections made at the beginning of fiscal year 2017—when the expected launch readiness date was October 2018—expected workforce levels to begin at 472 full-time equivalent staff and drop to 109 at the end of the fiscal year. However, technical challenges and delays in completing scheduled work did not allow for the planned workforce reduction and Northrop Grumman reported 496 full-time equivalent staff in September 2017, or 387 more than planned. According to JWST project officials and similar to previous years, Northrop Grumman's priority for fiscal year 2018 is to maintain schedule in order to ensure that the new launch window set from March to June 2019 can be met. As a result, Northrop Grumman's contractor workforce levels are expected to continue to be elevated through JWST's final integration and test phase in fiscal year 2019 where the spacecraft and OTIS will be integrated before shipment to the launch site.

Northrop Grumman submitted a cost overrun proposal to NASA in July 2016, primarily to address costs associated with sustaining its workforce at higher levels than planned in fiscal year 2017. An overrun proposal seeks to increase the value of a cost-reimbursement contract when the total estimated cost is less than the contract's estimated cost to complete the performance of the contract.[12]

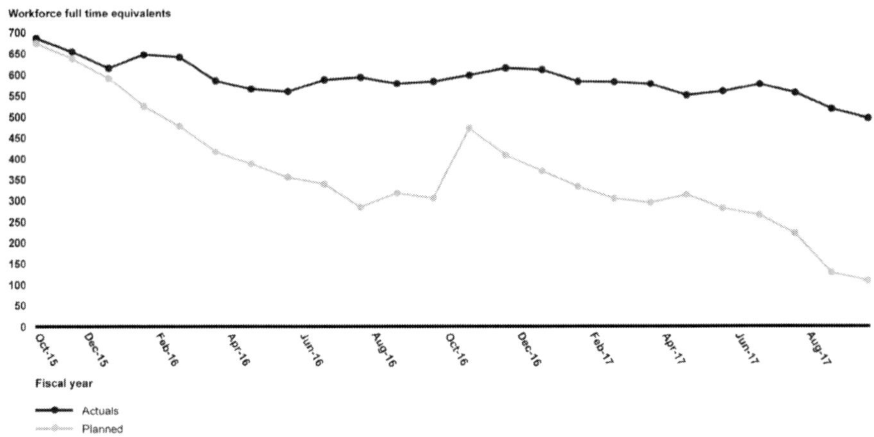

Source: GAO analysis of contractor data. | GAO-18-273.

Note: New projections were made at the start of each fiscal year. At the start of fiscal year 2017, in October 2016, new projections were made for the fiscal year resulting in a spike in projected workforce levels.

Figure 6. Northrop Grumman Workforce Levels in Fiscal Years 2016 and 2017, Projections at Beginning of Each Fiscal Year versus Actuals.

In addition to higher workforce levels, the overrun proposal replenished contractor management reserves that had been used to address technical issues, and addressed projected growth in the contractor's cost to complete work. NASA and the contractor completed negotiations in September 2017 and executed a contract modification that added $179.9 million to the value of the contract to cover Northrop Grumman's cost overrun and additional negotiated items, such as particle dampers.[14] This amount was intended to cover the cost of the remaining work through the expected launch date of October 2018. However, by September 2017

[12] JWST's contract with Northrop Grumman is a cost-plus-award-fee contract. Award fees are used when key elements of performance cannot be defined objectively, and, as such, require the project officials' judgment to assess contractor performance.

[14] Particle dampers provide a mechanism for diverting energy away from resonant structural vibrations of the observatory during launch.

Northrop Grumman had no remaining schedule reserves and a limited amount of cost reserves with which to address future costs.

Furthermore, the project determined—as discussed above—that the October 2018 launch window was not feasible and established a new launch window. According to JWST project officials, the project expects to issue a request for proposal in early 2018 to cover the costs for the remaining work through the new launch window. The project plans to use a significant portion of fiscal years 2018 and 2019 program cost reserves to address Northrop Grumman costs and unanticipated technical challenges. According to JWST program officials, if the contractor does not improve its schedule efficiency, the remaining reserves will be used to offset increased cost resulting from taking longer to complete the work.

For the sixth consecutive year, the JWST project managed spending within its allocated budget in fiscal year 2017. However, JWST is still resolving technical challenges and planned work continues to take longer to complete. Prudent management of its resources allowed the project to carry into fiscal year 2018 about a third more carry over funding than it had projected at the beginning of the fiscal year. Program officials said that assuming the remaining integration and tests proceed as planned and no long delays are encountered, the existing program resources accommodate the new launch window of March to June 2019. The project continues to identify funding options in the event of a delay of beyond the end of the launch window. Under the 2011 replan, Congress placed an $8 billion cap on formulation and development costs, but any long delays beyond the new launch window—which, as noted above, are likely— place the project at risk of exceeding this cap.

AGENCY COMMENTS AND OUR EVALUATION

We requested comments from NASA, but agency officials determined that no formal comments were necessary. NASA provided technical comments, which were incorporated as appropriate.

Cristina T. Chaplain
Director
Contracting and National Security Acquisitions

List of Committees

The Honorable Richard C. Shelby
Chairman

The Honorable Jeanne Shaheen
Ranking Member
Subcommittee on Commerce, Justice, Science, and Related Agencies
Committee on Appropriations
United States Senate

The Honorable John Culberson
Chairman

The Honorable José Serrano
Ranking Member
Subcommittee on Commerce, Justice, Science, and Related Agencies
Committee on Appropriations
House of Representatives

The Honorable Lamar Smith
Chairman

The Honorable Eddie Bernice Johnson
Ranking Member
Committee on Science, Space and Technology
House of Representatives

APPENDIX I: ELEMENTS AND MAJOR SUBSYSTEMS OF THE JAMES WEBB SPACE TELESCOPE (JWST) OBSERVATORY

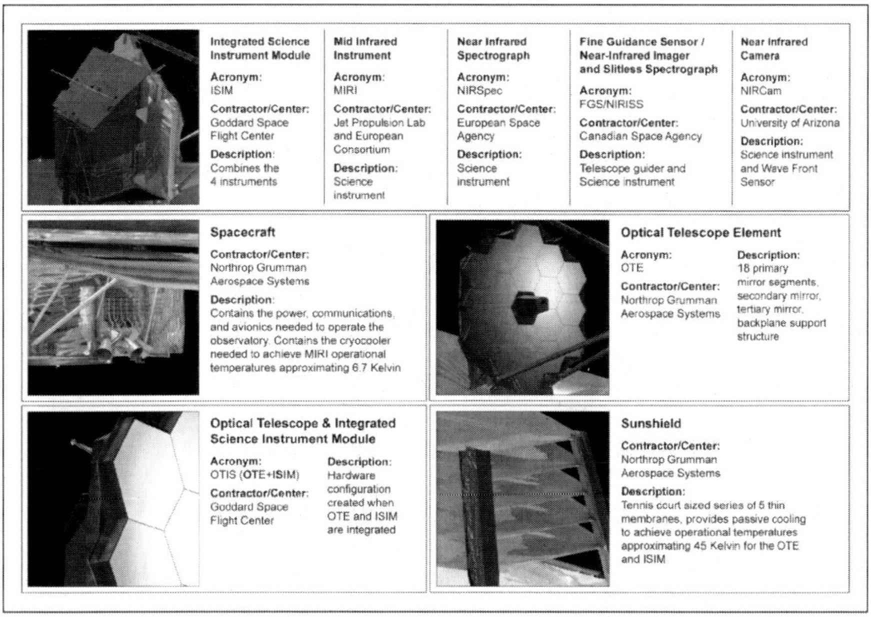

Souce: GAO (analysis); National Aeroautics and Space Administration (NASA) (date and images). | GAO-18-273.

Figure 7. Elements and Major Subsystems of the James Webb Space Telescope (JWST) Observatory.

APPENDIX II: STAFF ACKNOWLEDGMENTS

In addition to Cristina T. Chaplain, Richard Cederholm, (Assistant Director); Karen Richey, (Assistant Director); Jay Tallon, (Assistant Director); Brian Bothwell, Laura Greifner, Daniel Kuhn, Katherine Lenane, Jose Ramos, Carrie Rogers, Sylvia Schatz, and Roxanna Sun made key contributions to this chapter.

Appendix III: Accessible Data

Data Tables

Accessible Data for Figure 2: Integration and Test (I&T) Schedule for the James Webb Space Telescope

- July 2011 – March 2016
    - ISIM I&T: integrates and tests the 4 science instruments
- September 2014 – March 2016
    - OTE I&T: integrates and tests the optics and support structure
- February 2016 – July 2018
    - Spacecraft I&T: integrates the spacecraft and the sunshield
- April 2016 – April 2018
    - OTIS I&T: integrates the ISIM and the OTE
- July 2018 – March 2019
    - Observatory I&T: integrates spacecraft and OTIS
- April 2019 – June 2019
    - Shipping and launch site activities
- June 2019
    - Launch readiness date

Accessible Data for Figure 5: Schedule Reserve Held by Elements and Major Subsystems for the James Webb Space Telescope in 2016 and 2017

Elements and major subsystems	Schedule reserve (2016) in months	Current schedule reserve (2017) in months
Spacecraft	7	1.5
OTE + ISIM (OTIS)	6	4
Ground system	8.75	14.5

Accessible Data for Figure 6: Northrop Grumman Workforce Levels in Fiscal Years 2016 and 2017, Projections at Beginning of Each Fiscal Year versus Actuals

Date	Workforce full time equivalents (planned)	Workforce full time equivalents (actuals)
Oct-2015	674	686
Oct-2015	637	654
Dec-2015	591	615
Dec-2015	525	647
Feb-2016	477	641
Feb-2016	417	585
Apr-2016	387	565
Apr-2016	355	559
Jun-2016	339	587
Jun-2016	284	593
Aug-2016	317	578
Aug-2016	305	583
Oct-2016	472	598
Oct-2016	409	615
Dec-2016	370	611
Dec-2016	333	583
Feb-2017	304	582
Feb-2017	294	577
Apr-2017	313	550
Apr-2017	281	559
Jun-2017	265	576
Jun-2017	221	556
Aug-2017	128	518
Aug-2017	109	496

Accessible Data for Figure 7: Elements and Major Subsystems of the James Webb Space Telescope (JWST) Observatory

- Integrated Science Instrument Module
  - Acronym: ISIM
  - Contractor/Center: Goddard Space Flight Center
  - Description: Combines the 4 instruments
- Mid Infrared Instrument
  - Acronym: MIRI
  - Contractor/Center: Jet Propulsion Lab and European Consortium
  - Description: Science instrument
- Near Infrared Spectrograph
  - Acronym: NIRSpec
  - Contractor/Center: European Space Agency
  - Description: Science instrument
- Fine Guidance Sensor / Near-Infrared Imager and Slitless Spectrograph
  - Acronym: FGS/NIRISS
  - Contractor/Center: Canadian Space Agency
  - Description: Telescope guider and Science instrument
- Near Infrared Camera
  - Acronym: NIRCam
  - Contractor/Center: University of Arizona
  - Description: Science instrument and Wave Front Sensor
- Spacecraft
  - Contractor/Center: Northrop Grumman Aerospace Systems
  - Description: Contains the power, communications, and avionics needed to operate the observatory. Contains the cryocooler needed to achieve MIRI operational temperatures approximating 6.7 Kelvin
- Optical Telescope Element
  - Acronym: OTE
  - Contractor/Center: Northrop Grumman Aerospace Systems
  - Description: 18 primary mirror segments, secondary mirror, tertiary mirror, backplane support structure
- Optical Telescope & Integrated Science Instrument Module
  - Acronym: OTIS (OTE+ISIM)
  - Contractor/Center: Goddard Space Flight Center
  - Description: Hardware configuration created when OTE and ISIM are integrated

- Sunshield
    - Contractor/Center: Northrop Grumman Aerospace Systems
    - Description: Tennis court sized series of 5 thin membranes, provides passive cooling to achieve operational temperatures approximating 45 Kelvin for the OTE and ISIM

In: Commercial Crew Program …
Editor: Graeme N. Lutz
ISBN: 978-1-53614-980-7

Chapter 6

GLOBAL POSITIONING SYSTEM: BETTER PLANNING AND COORDINATION NEEDED TO IMPROVE PROSPECTS FOR FIELDING MODERNIZED CAPABILITY*

United States Government Accountability Office

WHY GAO DID THIS STUDY

GPS provides positioning, navigation, and timing data to civilian and military users who depend on this satellite-based system. Since 2000, DOD—led by the Air Force—has been working to modernize GPS and to keep the current system of satellites—known as the GPS constellation—operational, although these efforts have experienced cost and schedule growth.

---

* This is an edited, reformatted and augmented accessible version of the United States Government Accountability Office Report to Congressional Committees, Publication No. GAO-18-74, dated December 2017.

The National Defense Authorization Act for Fiscal Year 2016 contained a provision that the Air Force provide reports to GAO on GPS acquisition programs and that GAO brief the congressional defense committees. GAO briefed the committees in 2016 and 2017. This chapter summarizes and expands on information presented in those briefings.

This chapter assesses the extent to which DOD faces acquisition challenges (1) sustaining the GPS constellation; (2) developing a new ground control system; and (3) developing and fielding modernized receivers. GAO analyzed GPS quarterly acquisition reports and data, acquisition strategies, software and test plans, and other documents, and interviewed DOD and contractor officials.

What GAO Recommends

DOD should assign responsibility to an organization to collect test data, lessons learned, and design solutions so they may be shared. DOD concurred with the recommendation.

What GAO Found

The Department of Defense's (DOD) acquisition of the next generation Global Positioning System (GPS) satellites, known as GPS III, faces a number of acquisition challenges, but these challenges do not threaten DOD's ability to continue operating the current GPS system, which DOD refers to as the constellation, in the near term. Projections for how long the current constellation will be fully capable have increased by nearly 2 years to June 2021, affording some buffer to offset any additional satellite delays. While the first GPS III satellite has a known parts problem, six follow-on satellites—which do not—are currently scheduled to be launched by June 2021.

DOD is relying on a high-risk acquisition schedule to develop a new ground system, known as OCX, to control the broadcast of a modernized military GPS signal. OCX remains at risk for further delays and cost growth. To mitigate continuing delays to the new ground control system, the Air Force has begun a second new program—Military-code (M-code) Early Use—to deliver an interim, limited broadcast encrypted GPS signal for military use by modifying the current ground system. GAO will continue to monitor OCX progress.

DOD has made some progress on initial testing of the receiver cards needed to utilize the M-code signal. However, additional development is necessary to make M-code work with over 700 weapon systems that require it. DOD has begun initial planning for some weapon systems, but more remains to be done to understand the cost and schedule needed to transition to M-code receivers. The preliminary estimate for integrating and testing a fraction of the weapon systems that need the receiver cards is over $2.5 billion through fiscal year 2021 with only 28 fully and 72 partially funded (see figure). The cost will increase by billions when as yet unfunded weapon systems are included.

Source: GAO analysis of Department of Defense data. | Gao-18-74.

Status of Weapon Systems That Have Determined the Cost Needed to Transition to M-code Receivers through Fiscal Year 2021, as of February 2017

The level of development and procurement effort beyond the initial receiver cards is significant and will require close coordination across DOD. After the Air Force develops initial cards, the breadth and complexity of this acquisition will multiply, as the offices responsible for upgrading hundreds of weapon systems begin their own individual efforts to further develop and test the cards. However, DOD does not have an organization assigned to collect test data, lessons learned, and design solutions so that common design solutions are employed to avoid duplication of effort as multiple entities separately mature receiver cards. DOD therefore risks paying to repeatedly find design solutions to solve common problems because each program office is likely to undertake its own.

* * *

December 12, 2017
Congressional Committees

For more than two decades, the satellite-based Global Positioning System (GPS) has provided positioning, navigation, and timing (PNT) data to civilian and military users worldwide. Given its ubiquity throughout the world, failure, malfunction, or jamming of GPS can severely disrupt day-to-day civilian and military activities across the globe. The Department of Defense (DOD) maintains and operates GPS, which consists of three segments: orbiting satellites that continuously broadcast position and time data; a ground control system that operates the satellites and monitors and corrects signal data; and receivers used by civilians and the military in aircraft, ships, land vehicles, munitions, and hand-held units. Since 2000, the Air Force has led the multi-billion dollar effort to modernize GPS to sustain the satellite constellation—currently comprised of 31 satellites—by providing new civilian and military signals, enhancing cybersecurity, developing modernized military receivers, and countering known threats.

The National Defense Authorization Act for Fiscal Year 2016 contained a provision that the Air Force provide quarterly reports to GAO on the next generation GPS acquisition programs.[1] The Act also contained a provision that GAO brief congressional defense committees on the first

[1] Pub. L. No. 114-92, § 1621 (2015).

report, and at GAO's discretion, on subsequent quarterly reports. The Air Force delivered the first quarterly report to us on April 22, 2016. We assessed the report, briefed congressional committees in June 2016, and issued a report on that briefing.[2] During fiscal years 2016 and 2017, we briefed the congressional defense committees several additional times regarding the ongoing work. This chapter summarizes and expands on those briefings. This chapter assesses the extent to which DOD faces acquisition challenges (1) sustaining the GPS constellation; (2) developing a new ground system to control the broadcast of a modernized military GPS signal called military-code, or M-code; and (3) developing and fielding modernized receiver cards across the department.[3]

To conduct our work, we reviewed Air Force GPS fiscal year 2016 and 2017 quarterly acquisition reports, program acquisition baselines, integrated master schedules, acquisition strategies, software development plans, test plans, and other documents for five programs:

- GPS III,
- Next generation operational control system (OCX),
- Military GPS user equipment (MGUE),
- Contingency Operations (COps), and
- M-code Early Use (MCEU) programs.

We also interviewed officials from the GPS III, OCX, MGUE, COps, and MCEU programs; the prime contractors from all five programs; the Defense Contract Management Agency; and, where applicable, officials from DOD's Office of Cost Assessment and Program Evaluation; the Office of the Director, Operational Test and Evaluation (DOT&E); Under Secretary of Defense for Acquisition, Technology, and Logistics (USD AT&L); Joint Staff / J-6 Space Branch; Air Force Space Command; and

---

[2] GAO, Global Positioning System: Observations on Quarterly Reports from the Air Force, GAO-17-162R (Washington, D.C.: Oct. 17, 2016).

[3] M-code is a stronger, encrypted, military-specific GPS signal designed to meet military PNT needs.

each of the military services involved with the planning and procurement of the MGUE program. Appendix I contains a more detailed description of our scope and methodology.

We conducted this performance audit from February 2016 to December 2017 in accordance with generally accepted government auditing standards. Those standards require that we plan and perform the audit to obtain sufficient, appropriate evidence to provide a reasonable basis for our findings and conclusions based on our audit objectives. We believe that the evidence obtained provides a reasonable basis for our findings and conclusions based on our audit objectives.

BACKGROUND

Collectively, the ongoing GPS acquisition effort aims to (1) modernize and sustain the existing GPS capability and (2) enhance the current GPS system by adding an anti-jam, anti-spoof cybersecure M-code capability.[4] Figure 1 below shows how GPS satellites, ground control, and user equipment—in the form of receiver cards embedded in systems—function together as an operational system.

Modernizing and sustaining the current GPS broadcast capability requires launching new satellites to replace the existing satellites that are near the end of their intended operational life as well as developing a ground control system that can launch and control both existing and new satellites. Sustaining the current GPS broadcast capability is necessary to ensure the quality and availability of the existing broadcast signals for civilian and military GPS receivers. The ongoing modernization of GPS began with three programs: (1) GPS III satellites; (2) OCX to control the satellites; and (3) MGUE increment 1 (which develops initial receiver test

[4] Anti-jam capability blocks signal interference (jamming). Anti-spoof capability protects users against false signals that adversaries may employ to imitate friendly GPS systems (spoofing).

cards for military ships, ground vehicles, or aircraft).[5] Table 1 describes these programs.

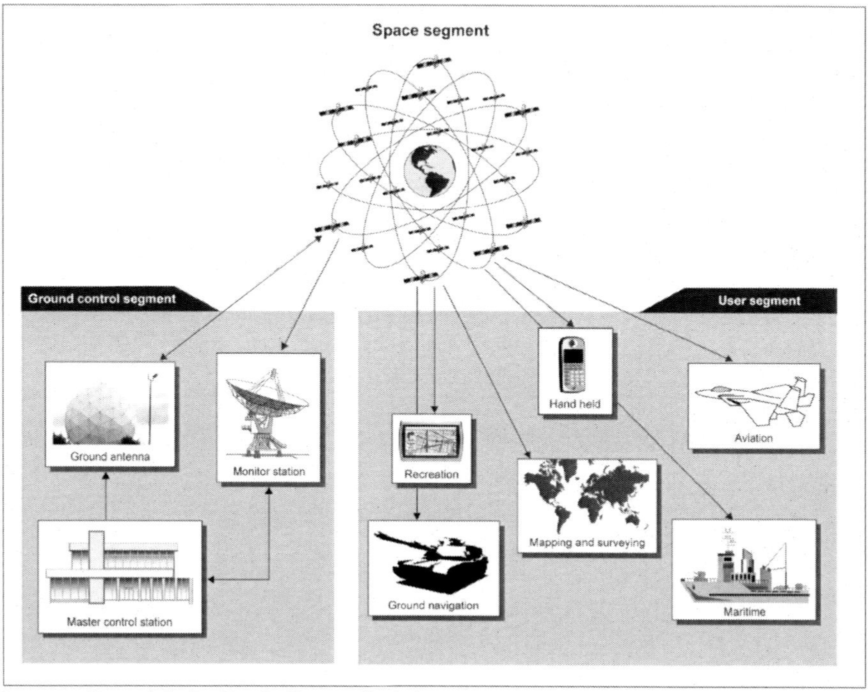

Source: Copyright© Corel Corp. all right reserved (map); Art Explosion (images); GAO analysis of Department of Defense data. | GAO-18-74.

Figure 1. GPS Operational System.

Delays to OCX of more than 5 years led the Air Force to create two additional programs in 2016 and 2017 to modify the current GPS ground system to control GPS III satellites and provide a limited M-code broadcast. As a result, there are currently five total GPS modernization programs. Table 2 provides a description of the two new programs.

[5] The MGUE increment 1 program is developing two receiver card types for (1) ground and (2) aviation/maritime applications on four military service-nominated "lead platforms," which serve as initial representatives for all platforms' integration.

Table 1. Current Global Positioning System Modernization Programs

	Program	Description
Space Segment	Global Positioning System (GPS) III	GPS III satellites will supplement and eventually replace the constellation of GPS satellites now in orbit, sustaining current capabilities and providing new signals. The GPS constellation currently consists of multiple satellite generations. Previous generations of satellites have been transmitting the military code—or M-code—signal for more than 11 years. GPS III satellites will provide a stronger M-code signal. Lockheed Martin was awarded the contract to build the first 10 GPS III satellites. The Air Force plans to competitively award satellites 11 through 32 in fiscal year 2018.
Ground Control Segment	Next Generation Operational Control System (OCX)	OCX will replace the current ground control system, known as the operational control segment (OCS). OCS lacks modern cybersecurity protections and cannot control–or enable–modernized features of the two latest generations of GPS satellites now in orbit, including M-code and some new civilian signals. Raytheon is the prime contractor. OCX is being developed in a series of blocks: 1 Block 0—will provide the launch and checkout system and supports initial testing of GPS III satellites. GPS III satellites cannot be launched without OCX block 0. It also provides most of the modern cybersecurity—a key advancement in securing the system. 2 Blocks 1 and 2—will provide command and control for previous generations of satellites and GPS III satellites, as well as monitoring and control for both current and modernized signals. Block 2 will provide the full M-code broadcast capability. The planned operational date for block 2 is January 2022.
User Segment	Military GPS User Equipment (MGUE)	MGUE is developing the M-code capable receiver cards with enhanced positioning, navigation, and timing (PNT) capabilities and improved resistance to existing and emerging threats, such as jamming. The Air Force is developing MGUE in two increments:

	Program	Description
		1. Increment 1—will provide initial ground, maritime, and aviation security-certified receiver cards. It entered system development in January 2017 and is scheduled to conduct operational testing on four military service-nominated lead platforms by April 2021. Three contractors—L3 Technologies, Raytheon, and Rockwell Collins—are building initial receiver cards. The cards consist of hardware—the physical card itself—and software—the coding within the card—that must be developed, tested, and integrated. 2. Increment 2—will provide more compact receiver cards to be used when size, weight, and power must be minimized, such as on handheld receivers, space receivers, and precision munitions. The Air Force plans to deliver the acquisition strategy for increment 2 in March 2018.

Source: GAO analysis of Department of Defense data. | GAO-18-74.

Table 2. Modifications to Current Operational Control Segment

Program	Description
Contingency Operations (COps)	COps will be a software modification to the existing operational control segment (OCS) started in 2016 to sustain the GPS constellation's capabilities. COps is planned to enable operational control of GPS III satellites and the legacy signals they will broadcast, while OCS will continue to control GPS II satellites and their signals. COps is a bridge capability needed to fill in the gap created by OCX development delays. Lockheed Martin is the prime contractor for this work and also built and maintains OCS.
M-code Early Use (MCEU)	MCEU will be an additional software modification to OCS after COps is completed. It will provide some operational control of specific M-code broadcast signals, but will lack the cybersecurity of OCX, among other features. This is a bridge capability needed to fill in the gap created by the OCX development delays. Lockheed Martin is the prime contractor.

Source: GAO analysis of Department of Defense data. | GAO-18-74.

Table 3. Schedule Challenges in Modernized GPS Programs

Program	Challenges
Global Positioning System (GPS) III	Since the program's original 2008 baseline, the first GPS III satellite's planned launch date has been delayed by four years—from April 2014 to May 2018. The Air Force and Lockheed Martin had repeated delays due to technical challenges with the navigation payload, resulting in a January 2016 cost and schedule rebaseline—or a recalculation by the program of new costs based on the new schedule.
Next Generation Operational Control System (OCX)	Since the program's original 2012 baseline: OCX block 1 ready to transition to operations date has been delayed by 63 months—from October 2016 to January 2022. The program's cost increases prompted a cost and schedule rebaseline in 2015 and declaration of a critical Nunn-McCurdy unit cost breach in 2016.[a] The Air Force and Raytheon have struggled with OCX development hurdles and unrealistic schedules with numerous root causes.
Military GPS User Equipment (MGUE) Increment 1	The Air Force began MGUE technology development in April 2012, but the cost and schedule baseline for the program was not approved until January 2017, over a year later than originally planned. The Air Force and the MGUE contractors—L3 Technologies, Raytheon, and Rockwell Collins—have experienced varying degrees of schedule delays, in part, due to lack of software engineering rigor and technology readiness shortcomings. The Air Force pursued an accelerated MGUE development and eliminated a key milestone based on early optimistic assessments of technology maturity. However, independent reviews highlighted concerns over development maturity and the characterization of demonstration test results.

Source: GAO analysis of Department of Defense data. | GAO-18-74.

[a] 10 U.S.C. § 2433, commonly referred to as Nunn-McCurdy, requires the Department of Defense to notify Congress whenever a major defense acquisition program's unit cost experiences cost growth that exceeds certain thresholds.

All of the original GPS modernization programs—GPS III, OCX, and MGUE—have experienced significant schedule growth during development. Table 3 outlines several schedule challenges in the modernized GPS programs.

We found in 2015 that unrealistic cost and schedule estimates of the new ground control system and receiver card development delays could pose significant risks to sustaining the GPS constellation and delivering M-code.[6] At that time, we also made five recommendations so that DOD would have the information necessary to make decisions on how best to improve GPS modernization and to mitigate risks to sustaining the GPS constellation. We made four OCX-specific recommendations targeted to identify underlying problems, establish a high confidence schedule and cost estimate, and improve management and oversight. For MGUE, we recommended the Air Force add a critical design review before committing resources to allow the military services to fully assess the maturity of the MGUE design before committing test and procurement resources. DOD concurred with the four recommendations on OCX and partially concurred on the MGUE recommendation. Since 2015, our annual assessment of DOD weapon systems has shown that some of the original GPS programs have continued to face cost or schedule challenges, increasing the collective cost to modernize GPS by billions of dollars.[7] Appendix III outlines the cost increases that have resulted.

Key GPS Modernization Points

According to our analysis, over the next decade or more, DOD plans to achieve three key GPS modernization points: (1) constellation sustainment, (2) M-code broadcast, and (3) M-code receivers fielded.

Figure 2 shows the current sequencing of the three points and the intervals when they are planned to be achieved, if known.

Throughout this chapter, we will use figures based on this one to highlight the separately-managed programs DOD plans to synchronize to

---

[6] GAO, GPS: Actions Needed to Address Ground System Development Problems and User Equipment Production Readiness, GAO-15-657 (Washington, D.C.: Sept. 9, 2015).

[7] GAO, Defense Acquisitions: Assessment of Selected Weapon Programs, GAO-17-333SP (Washington, D.C.: Mar. 30, 2017); Defense Acquisitions: Assessment of Selected Weapon Programs, GAO-16-329SP (Washington, D.C.: Mar. 31, 2016).

achieve each of the three identified modernization points. Some GPS capabilities require the delivery of more than one program, which must compete for limited resources, such as testing simulators. The Air Force coordinates the interdependent activities of the different programs and contractors in order to achieve each modernization point.

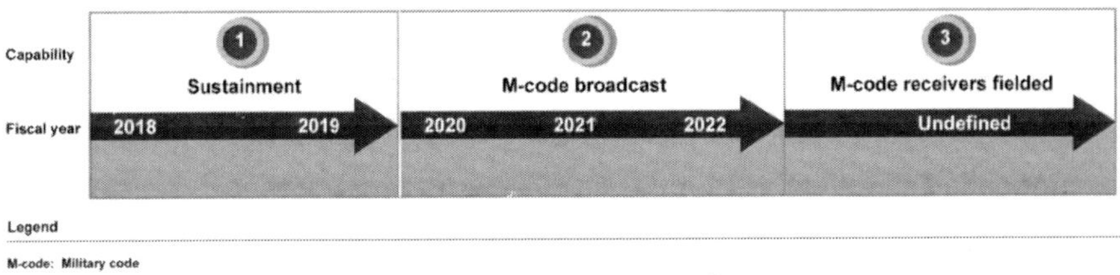

Source: GAO analysis of Department of Defense data. | Gao-18-74.

Note: This depiction reflects the current plan after numerous delays, not the original planned dates for all capabilities.

Figure 2. Current Key Global Positioning System Modernization Points.

GPS Satellite Constellation

The satellites in the GPS constellation broadcast encrypted military signals and unencrypted civilian signals and move in six orbital planes approximately 12,500 miles above the earth.

The GPS constellation availability performance standards commit the U.S. government to at least a 95 percent probability of maintaining a constellation of 24 operational GPS satellites to sustain the positioning services provided to both civilian and military GPS users. Therefore, while the minimum constellation consists of satellites occupying 24 orbital slots—4 slots in each of the six orbital planes—the constellation actually has 31 total satellites, generally with more than four in each plane to meet the 95 percent probability standard. These additional satellites are needed to provide uninterrupted availability in case a satellite fails. The constellation includes three generations of satellites with varying capabilities and design lives.

What is a Global Positioning System (GPS) satellite orbital plane and how many are there?

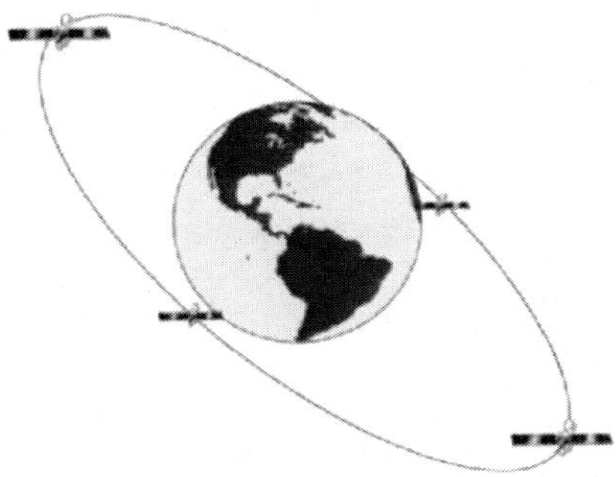

An orbital plane is an imaginary flat disc containing an Earth satellite's orbit. One orbital plane, as is shown above, represents the trajectory a GPS satellite follows as it circles the Earth in space.

The GPS constellation has six orbital planes. Each contains at least 4 satellites that allow the constellation to meet the minimum requirement of 24 satellites.

Source: Department of Defense (data); Art explosion (images). | GAO-18-74.

We found in 2010 and 2015 that GPS satellites have proven more reliable than expected, greatly exceeding their initially predicted life expectancies.[8] Nevertheless, the Air Force must regularly replace satellites to meet the availability standard, since operational satellites have a finite lifespan. Excluding random failures, the operational life of a GPS satellite tends to be limited by the amount of power that its solar arrays can produce.[9] This power level declines over time as the solar arrays degrade in the space environment until eventually they cannot produce enough power to maintain all of the satellite's subsystems. Consequently, the Air Force monitors the performance of operational satellites in

[8] GAO, Global Positioning System: Challenges in Sustaining and Upgrading Capabilities Persist, GAO-10-636 (Washington, D.C.: Sept. 15, 2010); GAO-15-657.

[9] The solar array is the component of the satellite's power subsystem upon which the solar cells are mounted.

order to calculate when new satellites need to be ready to join the constellation.

The 10 GPS III satellites currently under contract and in production with Lockheed Martin will provide a range of performance enhancements over prior GPS satellite generations. The GPS III satellites were designed to provide a longer life than previous generations, greater signal accuracy, and improved signal integrity—meaning that the user has greater assurance that the broadcast signal is correct. When they are eventually controlled through the OCX ground control system, the satellites will also offer a stronger M-code signal strength than prior GPS satellite generations. They will also include an additional civilian signal known as L1C, which will permit interoperability with European, Japanese, and other global navigation satellite systems for civilian users. Figure 3 describes the evolution of GPS satellite generations, including capabilities and life-span estimates.

Ground Control Segment

The current GPS ground control segment, OCS, primarily consists of software deployed at a master control station at Schriever Air Force Base, Colorado, and at an alternate master control station at Vandenberg Air Force Base, California. The ground control software is supported by 6 Air

Force and 11 National Geospatial-Intelligence Agency monitoring stations located around the globe along with four ground antennas that communicate with the moving satellites. Information from the monitoring stations is processed at the master control station to determine satellite clock and orbit status. As each of the three ground control segment programs—COps, MCEU, and OCX—is completed or partially completed, they will each introduce new capabilities, eventually culminating in the delivery of the full M-code broadcast planned for January 2022.

	GPS IIR	GPS IIR-M	GPS IIF	GPS III
		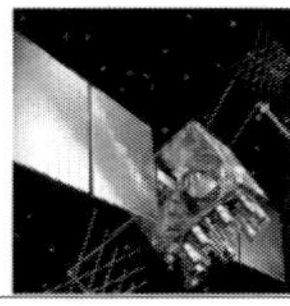	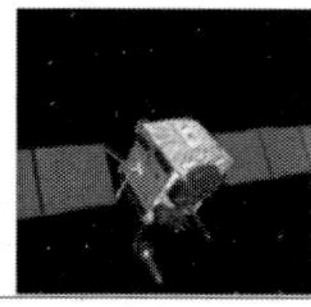	
Number	12 operational	7 operational	12 operational	10 under contract; additional 22 planned
Design life[a]	7.5 years	7.5 years	12 years	15 years
Mean life estimate[b]	26.2 years	18.9 years	12.5 years	16.3 years
Launched	1997 - 2004	2005 - 2009	2010 - 2016	First planned for May 2018
Capabilities	• Civilian and military signal broadcasts as provided by prior IIA generation satellites • Improved time keeping with on board clock monitoring	**IIR capabilities, plus:** • Second civilian signal • Second military signal (M-code) for enhanced jam resistance • Ability to increase signal power to improve resistance to jamming	**IIR-M capabilities, plus:** • Third civilian signal for transportation safety requirements • Improved accuracy, signal strength, and quality • Advanced atomic clocks	**IIF capabilities, plus:** • Fourth civilian signal to enable GPS interoperability with foreign satellite navigation systems • Enhanced signal reliability, accuracy, and integrity • Stronger military signal to improve jamming resistance

GPS Global Positioning System

Source: GAO analysis of Departmen of Defense information (data); U.S. government (images). | GAO-18-74.

Figure 3. Active and Future GPS Satellite Generations.

Receiver Cards

GPS receiver cards determine a user's position and time by calculating the distance from four or more satellites using the navigation signals on the satellites to determine the card's location. All warfighters currently acquire, train with, and use GPS receivers. Until MGUE receiver cards are developed and available for production, all DOD weapon systems that use GPS will continue to use the current GPS Selective Availability/Anti-Spoofing Module (SAASM) receiver card or an older version.[10] The Ike Skelton National Defense Authorization Act for Fiscal Year 2011 generally prohibits DOD from obligating or expending funds to procure GPS user equipment after fiscal year 2017 unless that equipment is capable

[10] For the purposes of this report, weapon systems include all DOD programs, platform systems, and subsystems that incorporate GPS.

of receiving M-code.[11] Under certain circumstances this requirement may be waived or certain exceptions may apply. The increment 1 receiver cards range in size from approximately 2 inches by 3 inches for the ground card up to 6 inches by 6 inches for the aviation/maritime card. Figure 4 below shows an illustration of a MGUE receiver card.

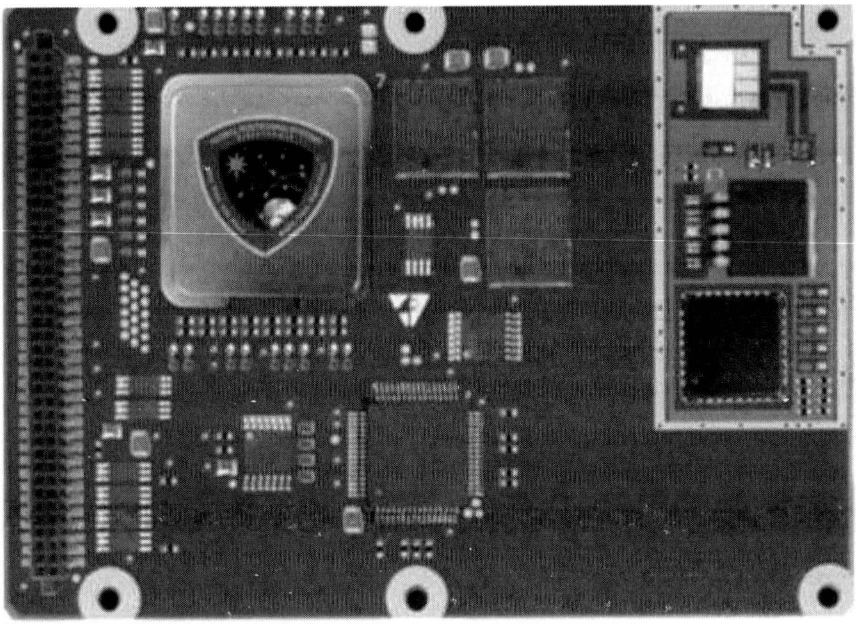

Source: Air Force. | GAO-18-74.

Figure 4. MGUE Receiver Card.

DOD has previously transitioned its weapon systems gradually from one generation of GPS receivers to the next. For example, some weapon systems have either upgraded or are still in the process of upgrading to the current SAASM receivers that were introduced in 2003, while others are still equipped with older cards. DOD anticipates that the length of time

[11] Pub. L. No. 111-383, § 913 (the Act) provides that none of the funds authorized to be appropriated or otherwise made available under the Act or any other act for DOD may be obligated or expended to purchase GPS user equipment during the fiscal years after fiscal year 2017 unless the equipment is capable of receiving M-code. This limitation does not apply to purchases of passenger or commercial vehicles in which GPS equipment is installed. Additionally, the Secretary of Defense may waive this limitation under certain circumstances.

necessary to transition to MGUE will require users to operate with a mix of receiver cards. Hundreds of different types of weapon systems require GPS receiver cards, including ships, aircraft, ground vehicles, missiles, munitions, and hand-held devices, among others, across all military services. The Air Force funds the MGUE program, providing funding to the military services so they can acquire, integrate, and operationally test the receiver cards on four service-specific lead platforms. These platforms are intended to test the card in the military services' ground, aviation, and maritime environments: (1) Army—Stryker ground combat vehicle; (2) Air Force—B-2 Spirit bomber; (3) Marine Corps—Joint Light Tactical Vehicle (JLTV); and (4) Navy—DDG-51 Arleigh Burke destroyer. Figure 5 depicts selected weapon systems that will need to install M-code capable receiver cards.

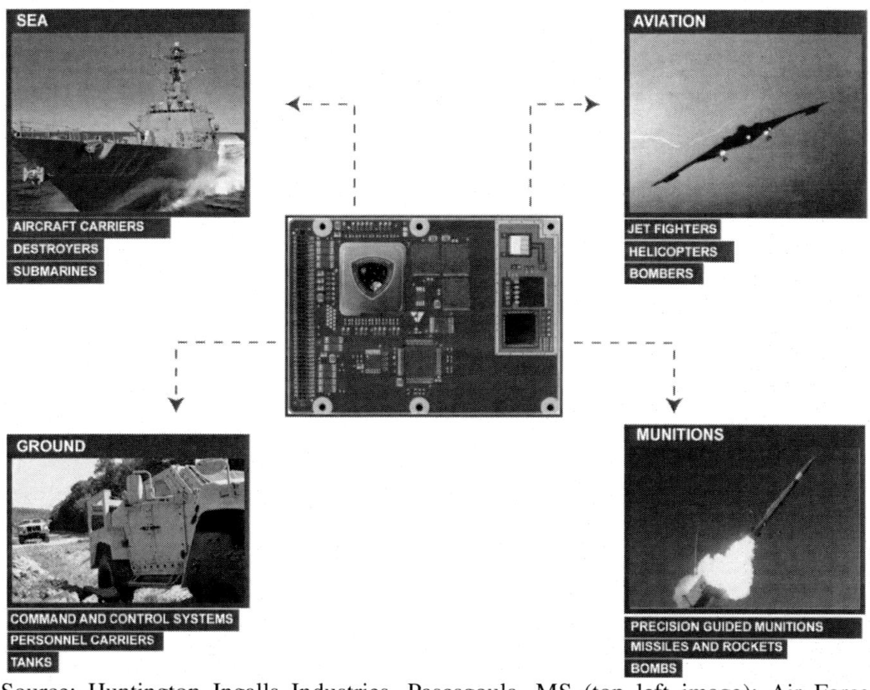

Source: Huntington Ingalls Industries, Pascsgoula, MS (top left image); Air Force (center and top right image); Army (bottom left and right image). | GAO-18-74.

Figure 5. Selected Weapon Systems That Require Military Code Receiver Cards.

ACQUISITION RISKS PERSIST ON GPS III SATELLITES BUT DO NOT THREATEN SUSTAINMENT OF THE CONSTELLATION IN THE SHORT TERM

The Air Force has made some progress toward ensuring continued constellation sustainment since our September 2015 report and should be able to sustain the current service because of the length of life of the current satellites. The current GPS constellation is now projected to meet its availability performance standard (in the absence of operational GPS III satellites) into June 2021—an increase of nearly 2 years over previous projections. This increase will give the Air Force more schedule buffer in the event of any additional delays to the GPS III satellite program.

However, the Air Force still faces technical risks and schedule pressures in both the short and long term. In the short term, schedule compression with the first GPS III satellite is placing the satellite's launch and operation at risk of further delays. In the long term, most of the satellites under contract will have been launched before operational testing is completed, limiting Air Force corrective options if issues are discovered. Figure 6 shows the schedule for programs that need to be delivered to modernize and sustain the GPS satellite constellation.

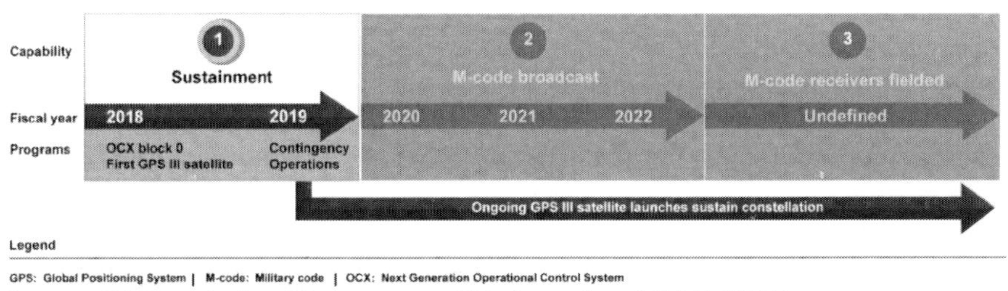

Source: GAO analysis of Department of Defense data. | GAO-18-74.

Figure 6. Schedule to Modernize and Sustain the GPS Satellite Constellation.

Progress: Programs Advancing to Support Constellation Sustainment Requirements

The Air Force has made progress since our last report in September 2015 on the three programs (GPS III, OCX, and COps) needed to support GPS constellation sustainment, readying both ground control and the satellite for the first GPS III satellite's launch, testing, and eventual operation. Raytheon delivered OCX block 0, the launch and checkout system for GPS III satellites, in September 2017. The Air Force took possession of OCX block 0 in October 2017 and will finally accept it at a later date after OCX block 1 is delivered.[12] Lockheed Martin completed the assembly, integration, and testing for the first GPS III satellite and in February 2017 the Air Force accepted delivery in advance of its currently scheduled May 2018 launch. As noted earlier, because of delays to OCX block 1, the Air Force initiated the COps program to ensure an interim means to control GPS III satellites. Without COps, no GPS III satellites can join the constellation to sustain it until OCX block 1 is operational in fiscal year 2022. In September 2016, COps formally started development, establishing a cost baseline of approximately $162 million to meet an April 2019 delivery. The COps program began software coding in November 2016, after a design review established that the product design would meet the Air Force's intended needs.

Short-Term Challenges: Compressed and Concurrent Schedules, Component Issues with the First GPS III Satellite

The Air Force continues to struggle with keeping multiple, highly compressed, interdependent, and concurrent program schedules synchronized in order to sustain and modernize the GPS constellation. Figure 7 shows some of the schedule challenges of the three programs needed for constellation sustainment and modernization.

---

[12] The Air Force took possession after inspection by signing a Certificate of Conformance.

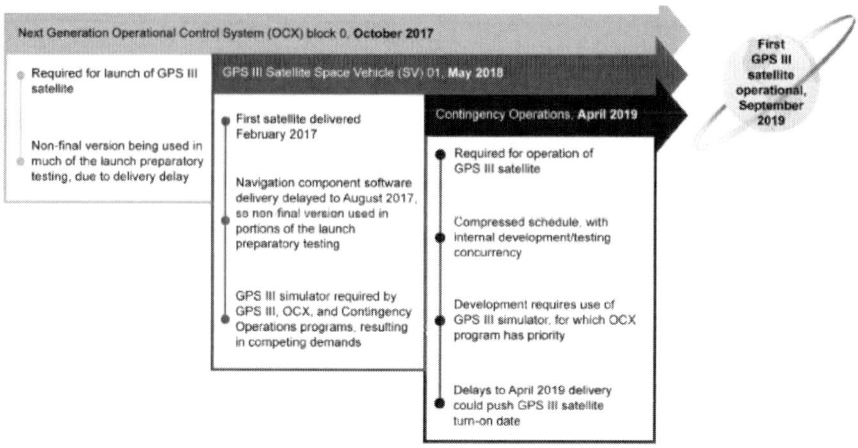

Source: GAO analysis of Department of Defense data. | GAO-18-74.

Figure 7. Global Positioning System (GPS) Challenges and Deliveries Needed to Launch the First GPS III Satellite.

Launching and operating the new GPS III satellite is a highly complex effort, since it requires synchronizing the development and testing schedules of OCX block 0, the first GPS III satellite, and the COps programs. For the Air Force to achieve its objective of making the first GPS III satellite operational by September 2019, numerous challenges (discussed below) must be addressed in the next 18 months on all three programs. If any of the three programs cannot resolve their challenges, the operation of the first GPS III satellite—and constellation sustainment—may be delayed.

OCX Block 0 and Pre-Launch Testing Schedules

With the goal of launching the first GPS III satellite in March 2018, the Air Force restructured its pre-launch integrated satellite and ground system testing in the summer of 2016, compressing the overall testing timeframe from 52 weeks to 42 weeks.[13] More OCX block 0 delays in

---

[13] Independent of OCX development, the Air Force moved the first GPS III target launch date from March to May 2018 in the spring of 2017. According to Air Force officials, the change was the result of a decision to launch the first GPS III satellite on a Falcon 9 rocket, rather than a Delta IV.

early fiscal year 2017 complicated Air Force test plans, resulting in changes to the sequence and timing of events, the introduction of concurrency at various points throughout the testing, the use of incomplete software in early testing, and an increase in the likelihood of discovering issues later in prelaunch integrated testing. Air Force officials stated that some pre-launch testing revisions streamlined the overall test plan since the merging of certain test events allowed multiple objectives to be met by the same event. Raytheon delivered OCX block 0, the launch and checkout system for GPS III satellites, in September 2017. The Air Force took possession of OCX block 0 in October 2017 and will finally accept it at a later date after OCX block 1 is delivered. However, if issues requiring corrective work are discovered during subsequent integrated testing, the GPS III launch schedule may be delayed further since there is minimal schedule margin on OCX block 0 for correcting any additional problems that may be found.

First GPS III Satellite Capacitors

There are hundreds of capacitors—devices used to store energy and release it as electrical power—installed in each GPS III satellite. In 2016, while investigating capacitor failures, the Air Force discovered that the subcontractor, then known as Exelis (now Harris Corporation), had not conducted required qualification testing for the capacitor's operational use in GPS III satellites. The Air Force conducted a review of the components over many months, delaying program progress while a subcontractor qualified the capacitor design as suitable for use on the GPS III satellite. However, the Air Force concluded that Harris Corporation failed to properly conduct a separate reliability test of the particular production lot from which the questionable capacitors originated. The Air Force directed the contractor to remove and replace the capacitors from that production lot from the second and third GPS III satellites. After weighing the technical data and cost and schedule considerations, the Air Force decided to accept the first satellite and launch it "as is" with the questionable capacitors installed.

Contingency Operations Schedule

The COps program is also pursuing a compressed and concurrent development and testing schedule to be operational as planned in September 2019. The COps acquisition strategy document acknowledges that the program's timeline is aggressive. DOT&E has highlighted the compressed COps schedule as a risk, since the limited time between the developmental and operational testing permits little time for the evaluation of test results and resolution of any deficiencies found.

The COps program has already begun drawing from its 60-day schedule margin, with a quarter of this margin used within the first 5 months after development started. According to Air Force officials, this margin use was the result of unplanned delays certifying a software coding lab. Additionally, the program schedule has concurrent development and testing, which in our previous work we have noted is often a high risk approach but is sometimes appropriate for software development.[14]

COps faces further schedule risk from its need for shared test assets, particularly the GPS III satellite simulator, a hardware- and software-based ground system that simulates GPS III function, which is also required by the GPS III and OCX programs. According to a DOT&E official, the OCX program receives priority over COps for the use of the GPS III satellite simulator, since the testing asset is heavily needed in the development of the ground control system.

Because of the competing demands for this resource, which Air Force and DOT&E officials maintain requires lengthy and complex software reconfigurations to repurpose the simulator from one test event to the next, the Air Force is using a less realistic and purely software-based simulator for the testing of COps, where possible.

---

[14] GAO, Missile Defense: Opportunity Exists to Strengthen Acquisitions by Reducing Concurrency, GAO-12-486 (Washington, D.C.: Apr. 20, 2012); GAO-16-329SP.

Short-Term Risk Mitigation: Nearly 2 Years of Schedule Buffer to When First GPS III Satellite Needed

Recent data show that the current satellites in the GPS constellation are expected to remain operational longer than previously projected, creating an additional, nearly 2-year schedule buffer before the first GPS III satellite needs to be operational to sustain the current GPS constellation capability. The Air Force projected that the first GPS III satellite needed to be operational by September 2019 based on 2014 satellite performance data. However, our analysis of the Air Force's more recent May 2016 GPS constellation performance data indicates that, in order to continue meeting the constellation availability performance standard without interruption, the operational need for the first GPS satellite is now June 2021.

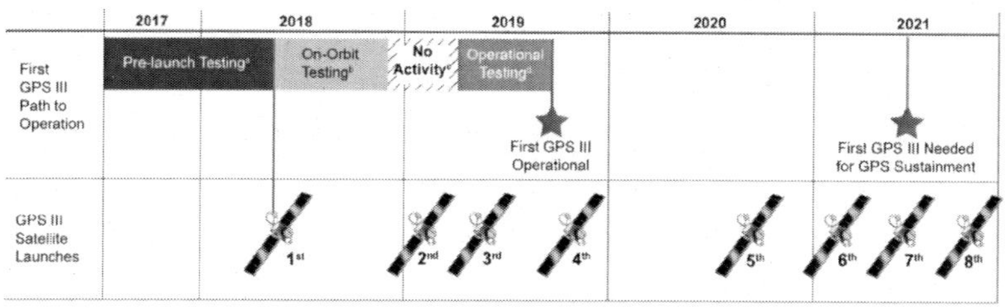

Source: GAO analysis of Department of Defense data. | GAO-18-74.

[a]Pre-launch testing entails operational-like testing of the OCX block 0 launch and checkout system and the GPS III satellite to demonstrate launch and satellite checkout capabilities prior to the May 2018 launch.

[b] On-orbit testing is the post-launch checkout testing of the GPS III satellite with OCX block 0 to provide early verification of some satellite subsystem functions.

[c] After completion of on-orbit testing, the GPS III satellite will orbit in a non-operational state until operational testing begins.

[d] Operational testing entails the testing of both the Contingency Operations (COps) ground control and the GPS III satellite to support an operational acceptance decision for both systems for non-modernized navigation signal capabilities.

Figure 8. Path to Global Positioning System (GPS) Constellation Sustainment.

This projection incorporates updated Air Force data from the current satellites that take into account an increase in solar array longevity expected for IIR and IIR-M satellites, according to Air Force officials. The

Air Force is likely to meet the constellation's June 2021 operational requirement because there are seven GPS III satellites planned to be launched by June 2021. Figure 8 shows the events leading to the launch and operation of the first GPS III satellite, achieving constellation sustainment once the first GPS III is operational, and subsequent GPS III launches that continue to support sustainment. The nearly 2-year buffer between planned operation and actual need for the first GPS III satellite permits the Air Force additional time to resolve any development issues. Because of this additional 2-year schedule buffer, we are not making a recommendation at this time to address the short term challenges we have identified but will continue to assess the progress of each of the programs and risks to constellation sustainment in our future work.

Long-Term Challenge: Most GPS III Satellites Under Contract Will Have Launched before Operational Testing Confirms Satellite Performance

The Air Force risks additional cost increases, schedule delays, and performance shortfalls because operational testing to confirm that GPS III satellites work as intended with OCX blocks 1 and 2 will not be completed until after the planned launch of 8 of the 10 GPS III satellites currently under contract. Due to delays to the OCX final delivery, the new ground control system will not be completed in time to control the GPS III satellites for the first few years they are in orbit (approximately 3.5 years). Consequently, GPS III operational testing will now occur in three phases—

1. in late fiscal year 2019 to confirm the satellites can perform similarly to the existing GPS satellites with COps;
2. in fiscal year 2020 to confirm the GPS III satellites can perform some of the new M-code capabilities with MCEU; and
3. in fiscal year 2022 to confirm the GPS III satellites can perform all of the new M-code capabilities with OCX blocks 1 and 2.

The first GPS III satellite is projected to complete operational testing of legacy signal capabilities in September 2019. By that point, the Air Force plans to have launched 3 of the 10 GPS III satellites, the fourth satellite is expected to be delivered, and major integration work will be underway on satellites 5 through 8. Therefore, if satellite shortcomings are discovered during any phase of the operational testing, the Air Force will be limited to addressing such issues through software corrections to satellites already on orbit. If any of the three phases of operational testing reveals issues, the Air Force may face the need for potentially costly contract modifications and delivery delays for satellites not yet launched. To offset this risk, the Air Force has obtained performance knowledge of GPS III satellites through ground testing of the first satellite, and findings from this testing have driven modifications to all ten satellites. Because of the rigor of the ground testing of the first satellite, Air Force officials maintain that the knowledge that might be obtained through on-orbit operational testing of the first GPS satellite would be minimal. However, a DOT&E official said that ground testing is limited to assessing system responses that are induced through the testing process and therefore may omit phenomena that might be experienced in actual system operation on orbit. We will continue to track the progress of operational testing in our future work.

Modernizing GPS Military Broadcast Challenged by High-Risk Development Schedules

DOD has established high-risk schedules for modernizing the GPS broadcast, or M-code signal, produced by GPS satellites. These risks are manifest in different ways. In the near term, the Air Force plans to provide a limited M-code broadcast—one that does not have all of the capabilities of OCX—in the MCEU program in fiscal year 2020. However, the MCEU schedule is high risk for its dependency on the timely completion of the COps program, for its aggressive schedule, and because of competition for limited test resources. Further, the full M-code broadcast capability, planned for fiscal year 2022, is at high risk of additional delays because (1) it is dependent on unproven efficiencies in software coding, (2) the program has not yet completed a baseline review, which may identify additional time needed to complete currently contracted work, and (3) there are known changes to the program that must be done that are not included in the proposed schedule.

High-Risk Programs Underlie Strategy to Deliver M-code Broadcast Capability

As noted above, the Air Force's plans for delivering the M-code broadcast involve two separate high-risk programs—MCEU and OCX blocks 1 and 2—delivered at separate times to make an operational M-code signal available to the warfighter. Figure 9 highlights the current forecasted operational schedules to deliver limited M-code broadcast capabilities with MCEU and full M-code broadcast with OCX.

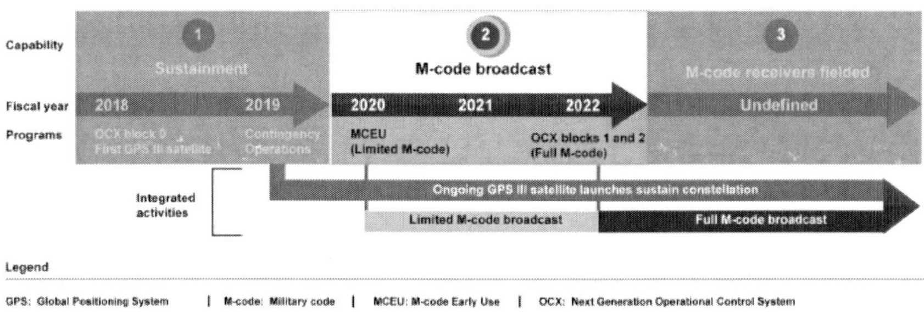

Source: GAO analysis of Department of Defense data. | GAO-18-74.

Figure 9. Schedule to Achieve Limited and Full M-code Broadcast.

MCEU

The MCEU program, created because of multiple delays to OCX and to partially address that program's remaining schedule risk, is itself a high-risk program that is dependent on the timely development of COps. Estimated to cost approximately $120 million, MCEU formally entered the acquisition process in January 2017 as a software-specific program to modify OCS. To develop MCEU, Lockheed Martin officials stated they will leverage personnel with expertise maintaining and upgrading OCS as well as utilize the staff working on COps. With a planned December 2019 delivery for testing and a September 2020 target to begin operations, the MCEU program faces several schedule risks. The Air Force's proposed plan anticipates a compressed software development effort, which the Air Force describes as aggressive. The Air Force has also identified potential risks to the MCEU schedule from competing demands by GPS III, OCX, COps, and MCEU for shared test resources. Air Force officials specifically noted competing demands for the GPS III simulator test resource. If development or testing issues arise in these other programs, those issues could delay the availability of the satellite simulator and thereby disrupt the planned MCEU development effort. According to program officials, the Air Force is working to mitigate this threat to the MCEU program through the use of a software-based simulator, when possible. Additionally, MCEU software development work is dependent on the timely conclusion of the COps effort—which, as previously mentioned, itself has an aggressive schedule and faces competition for a limited test resource. Air Force program officials have said that some Lockheed Martin staff planned to support MCEU will need to transfer from the COps effort. However, after reviewing the staffing plans at the MCEU contractor kickoff, Air Force officials said this is no longer viewed as a significant risk.

OCX Blocks 1 and 2

Raytheon has made some progress starting coding for OCX block 1 and taken the first steps toward implementing and demonstrating initial software development efficiencies that may benefit development for OCX blocks 1 and 2. The software efficiencies are built up in seven phases and need to be completed before the development process reaches each of the phases to take full advantage of the efficiencies they will create. Once ready, the efficiencies are inserted at different points in the software development schedule. For example, as of August 2017, the first of seven phases implementing the software development improvements was nearly complete, while the second phase was approximately two-thirds complete. Both are needed in place for insertion when the next phase of coding begins.

Further, the Air Force proposed a new rebaselined schedule in June 2017 as the final step to getting the program back on track after declaring a critical Nunn-McCurdy unit cost breach in 2016 when the program exceeded the original baseline by more than 50 percent. A Nunn-McCurdy unit cost breach classified as critical is the most serious type of breach and requires a program to be terminated unless the Secretary of Defense submits a written certification to Congress that, among other things, the new estimate of the program's cost is reasonable and takes other actions, including restructuring the program. In October 2016, DOD recertified the program, with a 24-month schedule extension. Under this newer proposed schedule Raytheon forecasts delivering blocks 1 and 2 in December 2020 with 6 months of extra schedule—a 30-month schedule extension—to account for unknown technical issues before OCX blocks 1 and 2 are due to the Air Force in June 2021. The Air Force projects operating OCX in fiscal year 2022 after completing 7 months of operational testing post-delivery.

Three factors place delivery of OCX blocks 1 and 2 in June 2021 at high risk for additional schedule delays and cost increases:

- First, the newly proposed June 2017 rebaselined schedule assumes significant improvements in the speed of software coding and testing that have not yet been proven, but will be introduced at various periods as software development proceeds. Whether Raytheon can achieve the majority of these efficiencies will not be known until the end of fiscal year 2018. However, the Defense Contract Management Agency, which independently oversees Raytheon's work developing OCX, noted in July 2017 a number of risks to the schedule, including that some initial assumed efficiencies had not been demonstrated. Specifically, they noted for initial coding on block 1 that Raytheon had achieved only 60 percent of the software integration maturity planned to that point in time in conjunction with greater numbers of software deficiencies that will require more time than planned to resolve.

- Second, the proposed rebaseline schedule has not yet undergone an integrated baseline review (IBR) to verify all of the work that needs to be done is incorporated into that schedule. The IBR is a best practice required by the Office of Management and Budget on programs with earned value management.[15] An IBR ensures a mutual understanding between the government and the contractor of the technical scope, schedule, and resources needed to complete the work. We have found that too often, programs overrun costs and schedule because estimates fail to account for the full technical definition, unexpected changes, and risks.[16] According to prior plans, the IBR would have taken place in early 2017, but it has been delayed multiple times for a number of reasons. A significant and recurring root cause of delays on the OCX program has been a lack of mutual understanding of the work between the Air Force and Raytheon. The IBR start was scheduled for November 2017 with completion in February 2018. Once conducted, the review may identify additional work not in the proposed schedule that needs to be completed before delivery. For example, Raytheon is conducting a review of hardware and software obsolescence. If significant additional obsolescence issues are found that need to be resolved before OCX blocks 1 and 2are delivered, the projected delivery date may need to be delayed further at additional cost.

---

[15] Earned value management is a project management tool that integrates the technical scope of work with schedule and cost elements for investment planning and control. It compares the value of work accomplished in a given period with the value of the work expected in that period. Differences in expectations are measured in both cost and schedule variances. The Office of Management and Budget requires agencies to use earned value management in their performance-based management systems for major acquisitions with developmental effort. Office of Management and Budget, Capital Programming Guide v. 3.0, Supplement to Office of Management and Budget Circular No. A-11, Preparation, Submission, and Execution of the Budget, para. I.5.5.4 & I.5.5.5 (July 2017).

[16] GAO, Cost Estimating and Assessment Guide: Best Practices for Developing and Managing Capital Program Costs, GAO-09-3SP (Washington, D.C.: Mar. 2009).

- Third, the OCX contract will likely be modified because the Air Force needs to incorporate into its contract with Raytheon a number of changes that are not currently a part of the proposed schedule. According to Air Force and contractor officials, negotiations are under way to determine which of these changes will be incorporated before OCX blocks 1 and 2 are delivered and which may be added after delivery. Air Force officials said that the incorporation of changes should be completed by February 2018.

Schedule risk assessments for OCX blocks 1 and 2 delivery vary, making it unclear when the full M-code broadcast will finally be operational. Government assessments of Raytheon's performance continue to indicate more schedule delays are likely. Table 4 shows the varying assessments of potential schedule delays by the Defense Contract Management Agency and the Air Force to the proposed June 2021 delivery date and the subsequent operational date that occurs 7 months later.

Table 4. Forecasts for the Next Generation Operation Control System (OCX) Blocks 1 and 2 Delivery with Subsequent Operational Date, as of July 2017

Forecaster	Schedule risk to current delivery of June 2021	Forecast delivery with schedule risk included	Operational date[a]
Defense Contract Management Agency	14 months	August 2022	March 2023
Air Force	6-7 months	January/February 2022	August/September 2022

Source: GAO analysis of Department of Defense data. | GAO-18-74.

[a] Air Force estimates that after delivery from Raytheon, the government will need an additional 7 months to test and prepare OCX to be operational.

In 2015, we made four recommendations to the Secretary of Defense, one of which was to use outside experts to help identify all underlying problems on OCX and develop high confidence cost and schedule estimates, among others, in order to provide information necessary to make decisions and improve the likelihood of success.[17] To date, none of these recommendations have been fully implemented but DOD has taken steps to address some of them. Further, because the Air Force has undertaken the COps and MCEU programs to provide interim capabilities to mitigate OCX delays for the full broadcast capability, we are not making additional recommendations at this time but will continue to monitor progress and risks to the acquisition of OCX.

Greater Coordination Needed to Prevent Duplication of Effort Developing and Fielding M-code Receivers

While technology development for the M-code receiver cards is underway, DOD has developed preliminary—but incomplete—plans to fully develop and field M-code receiver cards across the more than 700 weapon systems that will need to make the transition from the current technology. DOD has prepared initial cost and schedule estimates for department-wide fielding for a fraction of these weapon systems. While the full cost remains unknown, it is likely to be many billions of dollars greater than the $2.5 billion identified through fiscal year 2021 because there is significant work remaining to verify the initial cards work as planned and to develop them further after the MGUE increment 1 program ends. Without greater coordination of integration test results, lessons learned, and design solutions DOD is at risk of duplicated development work as multiple weapon system programs separately mature and field similar technologies on their own. Further, with the full M-code broadcast available in fiscal year 2022, a gap—the extent of which is unknown—between operationally broadcasting and receiving M-code exists. Figure 10 highlights the gap between the time the M-code signal will be operational and the undefined time M-code can be used by the military services.

---

[17] GAO-15-657.

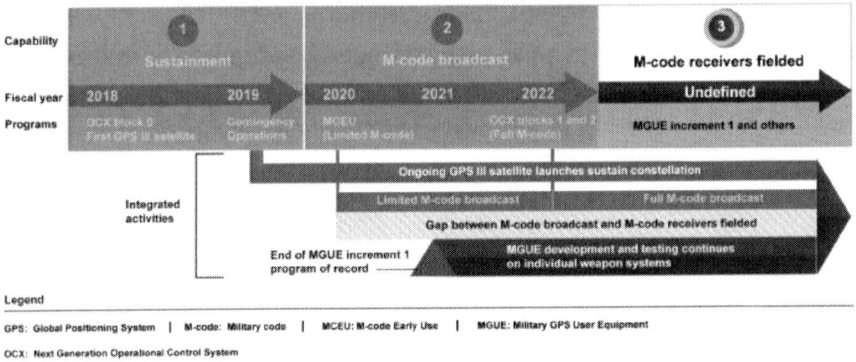

Source: GAO analysis of Department of Defense data. | Gao-18-74.

Figure 10. Schedule for Fielding Military Code Receiver Cards.

DOD Has Made Some Progress in Developing Technology for New M-code Receiver Cards

The Air Force program to develop initial M-code receiver test cards has made progress by establishing an acquisition strategy for this effort and maturing receiver test cards. In January 2017, DOD approved the MGUE increment 1 program to formally begin development, and it defined the criteria to end the program as (1) verifying technical requirements on all types of final receiver test cards; (2) certifying readiness for operational testing by the Air Force Program Executive Officer; (3) completing operational testing for the four lead platforms for, at a minimum, at least the first card available; and (4) completing manufacturing readiness assessments for all three contractors.

Within the MGUE increment 1 program, contractors are making progress toward delivering final hardware test cards and incremental software capabilities. For example, one contractor has achieved its initial security certification from the Air Force, which is a key step toward making the MGUE increment 1 receiver test card available for continued development and eventual procurement. Further, the MGUE increment 1 program is also conducting risk reduction testing in preparation for formal developmental verification testing, an important step that ensures the receiver cards meet technical requirements. Programs throughout DOD can make risk-based decisions to develop and test the receiver test cards after technical verification of the card's hardware and software. According to MGUE program officials, this is significant because it allows non-lead platforms to obtain and work with the cards sooner than the end date of operational testing on lead platforms.

Significant Development Work Remains to Eventually Field M-code Receiver Cards

Although the Air Force has made progress in maturing receiver test cards, significant development work remains to reach the point where the cards can ultimately be fielded on over 700 different weapon systems. For example, for MGUE increment 1, the Air Force must define additional technical requirements in order for the M-code receiver cards to be compatible and communicate with existing weapon systems. The Air Force will also need to conduct operational tests for each of the lead platforms—the Stryker ground combat vehicle; B-2 Spirit bomber; JLTV; and DDG-51 Arleigh Burke destroyer—before the full M-code signal is available with OCX. Because these tests will instead be conducted with the limited signal provided by MCEU, DOD risks discovering issues several years later once full operational testing is conducted. Further, according to military service officials and assessments by DOT&E, this operational testing will only be minimally applicable to other weapon systems because those other weapon systems have different operational requirements and integration challenges than the four lead platforms. As a result, additional development and testing will be necessary on an undetermined number of the remaining weapon systems to ensure the receiver cards address each system's unique interfaces and requirements. In 2018, DOD will also formally begin development for MGUE increment 2. Increment 2 will provide more compact receiver cards to be used when size, weight, and power must be minimized, such as on handheld receivers, space receivers, and munitions where increment 1 receiver cards are too large to work.

The military services are working to mitigate some of these development challenges. For example, Army officials told us they do not plan to field MGUE receiver cards on its lead platform, the Stryker, due to ongoing gaps in technical requirements. In addition, there is not a lead platform to demonstrate increment 1 on munitions since munition requirements were planned to be addressed in increment 2. However, to address its needs, the Army has initiated efforts to modify the MGUE increment 1 receiver card for some munitions that would otherwise need to wait for MGUE increment 2 technologies. Individual munition program offices within other military services have begun to do so as well. According to military service officials from the Army, Navy, and Marine Corps, it is essential that user needs are met by increment 2, or they will have to conduct additional development and testing. The Army previously identified gaps in increment 1 that the Air Force has either addressed in increment 1, has deferred to increment 2, or will need to be addressed outside of the MGUE increment 1 and 2 programs. Army and Navy officials also stated that they were concerned that any disagreements in requirements for increment 2 could lead to further fielding delays.

Finally, the transition from existing GPS receiver cards to M-code receiver cards is likely to take many years. We recently reported that transitioning all DOD platforms to the next generation of receiver cards will likely take more than a decade.[18] A lengthy transition has happened before, as previous efforts to modernize GPS to the current receiver cards, begun in 2003, are still underway and the older receiver cards are still being used. As a result, DOD anticipates that warfighters will have to operate with a mix of older and newer receiver cards.

---

[18] GAO, Space Acquisitions: DOD Continues to Face Challenges of Delayed Delivery of Critical Space Capabilities and Fragmented Leadership, GAO-17-619T (Washington, D.C.: May 17, 2017).

DOD Has Begun Cost and Schedule Planning; Full Cost Is Unknown but Likely to be Many Billions of Dollars

DOD has begun collecting preliminary information on M-code requirements for individual weapon systems. In December 2016, the USD AT&L directed the military services, the Missile Defense Agency (MDA), and Special Operations Command (SOCOM) to submit implementation plans with M-code investment priorities across weapon systems and munitions, including projected costs and schedules.[19] According to DOD, these M-code implementation plans are intended to provide DOD with a management and oversight tool for the fielding effort. In February 2017, each organization submitted its own implementation plan to USD AT&L. These plans were then briefed to the PNT Executive Management Board and PNT Oversight Council in February and March, respectively.[20]

However, these implementation plans are preliminary and based on assumptions about the Air Force's ability to achieve MGUE increment 1 and 2 technical requirements, the timeline required to do so, and the amount of development and test work that will remain for the receiver cards to be ready for production and fielding after the programs end. Since the MGUE increment 2 program has not started development, it has not yet finalized requirements. Once approved, the increment 2 program office will produce an acquisition strategy, schedule, and cost estimate. However, after the MGUE increment 2 program ends there is no detailed plan for completing development, testing, and fielding of M-code receiver cards for weapon systems across the department.

---

[19] The National Defense Authorization Act for Fiscal Year 2017 eliminated the position of USD AT&L effective February 1, 2018. The position will be divided into the Under Secretary of Defense for Research and Engineering and the Under Secretary for Defense for Acquisition and Sustainment. Pub. L. No. 114-328, § 901(a) and (b) (2016) (codified at 10 U.S.C. §§ 133a and 133b).

[20] The PNT Executive Management Board is chaired by the DOD Chief Information Officer and oversees the governance process in support of the PNT Oversight Council, including developing topics and recommendations for PNT Oversight Council actions. The PNT Oversight Council is co-chaired by USD AT&L and the Vice Chairman Joint Staff. The Council serves as the principal unified and integrated DOD governance body that ensures the DOD PNT Enterprise functions meet national objectives, is consistent with national policy and guidance, and that the mutually supporting systems continue to evolve to address emerging threats.

DOD has preliminary cost and schedule estimates for some weapon programs, but lacks a total cost at this point because the department does not include all efforts initiated by programs to meet specific needs, including those outside the MGUE increment 1 and 2 programs. The initial M-code implementation plans responded to what was requested but do not individually identify what the total cost will be for each organization to develop and field M-code receiver cards, so a total cost can be determined across DOD. Because USD AT&L required that the implementation plans include funding and schedule estimates for 2 to 3 years while directing that plans be resubmitted, at a minimum, every 2 years, weapon systems that will need M-code but were not considered an immediate priority were not included in the initial submissions. In addition, the military services, MDA, and SOCOM provided only initial cost estimates.[21] According to military service officials, these estimates were based on the current MGUE increment 1 program schedule and technical development and include risk-based decisions to partially fund specific programs until the MGUE increment 1 program matures. According to a USD AT&L official, the plans would both facilitate M-code implementation planning for the department and inform the issuance of waivers.[22] The official stated that as the acquisition programs critical to providing M-code capability mature, future implementations plans should provide more comprehensive estimates of cost and schedule to achieve M-code implementation for the department.

Our analysis of the M-code receiver card implementation plans found that initial funding estimates indicate a cost of over $2.5 billion to integrate and procure M-code receiver cards on only a small number of weapon systems out of the hundreds of types that need M-code receiver cards. The

---

[21] When we refer to "funded" or "partially funded" programs, we mean programs which DOD has planned—in full or in part—to fund in DOD's future years budget projections through fiscal year 2021, as reflected in the implementation plans.

[22] Pub. L. No. 111-383, § 913 (the Act) provides that none of the funds authorized to be appropriated or otherwise made available under the Act or any other act for DOD may be obligated or expended to purchase GPS user equipment during the fiscal years after fiscal year 2017 unless the equipment is capable of receiving M-code. This limitation does not apply to purchases of passenger or commercial vehicles in which GPS equipment is installed. Additionally, the Secretary of Defense may waive this limitation under certain circumstances.

full cost will be much larger—likely many billions of dollars because the majority of the weapon systems that need M-code receiver cards are not funded yet or are only partially funded, according to the M-code implementation plans. Specifically, the military services, MDA, and SOCOM identified 716 types of weapon systems in their February 2017 implementation plans that require almost a million M-code receiver cards. For example, the JLTV fleet—which provides protection for passengers against current and future battlefield threats for multiple military services—is one type of weapon system that will eventually need almost 25,000 receiver cards. Of the 716 types of weapon systems that will need M-code receiver cards, only 28—or less than 4 percent—are fully funded through fiscal year 2021. The remainder have either partially funded M-code development and integration efforts (72 weapon systems), or do not yet have funding planned (616 weapon systems). Additionally, the preliminary estimates to develop and procure M-code receivers on selected weapon systems do not all include funding beyond fiscal year 2021 that will be needed for further development, integration, and procurement. This means that DOD and Congress do not have visibility into how much additional funding could be needed to fully fund the remaining 96 percent of all weapon systems that need M-code receivers. Figure 11 shows the M-code development and integration efforts that are funded, partially funded, or unfunded through fiscal year 2021 across DOD weapon systems that will need M-code receiver cards.

Source: GAO analysis of Department of Defense data. | Gao-18-74.

Figure 11. Status of Weapon Systems That Have Determined the Cost Needed to Transition to M-code Receivers through Fiscal Year 2021, as of February 2017.

Because the implementation plans are a first step toward providing DOD leadership insight on this large set of acquisitions and they will be updated at least every 2 years by the different organizations within DOD, we are not making a recommendation at this time. However, we will continue to monitor DOD's cost and schedule planning.

DOD Risks Duplication of Effort Integrating and Testing M-code Receiver Cards

The level of development and procurement effort beyond MGUE increments 1 and 2 is significant and will require close coordination among the military services, MDA, and SOCOM. While Joint Staff officials stated that the DOD Chief Information Officer is working with the military services and Joint Staff to produce a user equipment roadmap to help guide that coordination, they said that these efforts are not yet complete. DOD has designated the Air Force to lead initial development of both larger and smaller test cards that other organizations will need to develop further to meet their individual needs. After the Air Force develops initial cards for both sizes, the breadth and complexity of this acquisition will multiply, as the offices responsible for upgrading hundreds of weapon systems begin their own individual efforts to further develop and test the cards so they work for the unique needs of their specific system. While some common solutions are being developed, Air Force officials said the military services and individual weapon systems will have the freedom to go to the contractors and begin their own development efforts.

DOD does not have a developed plan in place to help ensure that common design solutions are employed and that DOD avoids duplication of effort as multiple entities separately mature receiver cards. We previously found that duplication occurs when two or more agencies or programs are engaged in the same activities.[23] In this case, because the

---

[23] GAO, 2017 Annual Report: Additional Opportunities to Reduce Fragmentation, Overlap, and Duplication and Achieve Other Financial Benefits, GAO-17-491SP, (Washington, D.C.: Apr. 26, 2017).

individual organizations and program offices are likely to be pursuing individual and uncoordinated receiver card programs at different times with different contractors, DOD is at risk for significant duplication of effort. We previously found that establishing formal mechanisms for coordination and information sharing across DOD programs reduces the risk of gaps and results in more efficient and more effective use of resources.[24] Internal control standards also state that establishing clear responsibilities and roles in achieving objectives is key for effective management.[25] Further, DOD previously reported clear leadership ensures that programs and stakeholders are aligned with common goals.[26]

According to MGUE program officials, the MGUE increment 1 program is already capturing all issues observed in receiver test card risk reduction testing and sharing this information through a joint reporting system. However, while non-lead platforms may also report deficiencies in this system, there is no requirement that they do so, nor is there an entity responsible for ensuring data from testing, design, and development is shared between programs. We previously found that the absence of a formal process for coordination results in the potential for duplication, overlap, and fragmentation.[27] DOD therefore risks paying to repeatedly find design solutions to solve common problems because each program office is likely to undertake its own uncoordinated development effort. Some duplicated effort may already be occurring. Air Force officials have expressed concern that work is already being duplicated across the military services in developing embedded GPS systems to be integrated into aircraft. According to multiple DOT&E assessments, the absence of a plan across the wide variety of intended interfaces leaves significant risk in integrating the receiver cards, and therefore fielding cost and schedule risk for DOD.

---

[24] GAO-17-491SP.

[25] GAO, *Standards for Internal Control in the Federal Government,* GAO-14-704G, (Washington, D.C.: Sept. 2014).

[26] Defense Science Board, *Creating a DOD Strategic Acquisition Platform,* April 2009.

[27] GAO-17-491SP.

Conclusion

GPS is a national asset for civilians and the military service members who depend upon it each day. Any disruption to the system would have severe economic and military consequences. In keeping the system sustained and modernizing it with additional capabilities, DOD has spent billions of dollars more than planned developing five interdependent GPS programs. Developing these technologies is complex work with the collective effort already years behind initial estimates to provide the warfighter with a means to counter known threats, such as jamming, to the current system. It will be many years before M-code receiver cards are fielded at a cost that remains unknown but that will be substantially higher than the estimated $2.5 billion already identified through fiscal year 2021. In the short term, it is unclear when there will be a receiver card ready for production after the end of operational testing, and in the long term DOD risks wasting resources duplicating development efforts on weapon systems with similar requirements. Without better coordination of this effort, DOD risks unnecessary cost increases and schedule delays because there is no established process or place for collecting and sharing development and integration practices and solutions between programs.

Recommendations for Executive Action

We are making the following recommendation to DOD:

The Secretary of Defense should ensure that the Under Secretary of Defense for Acquisition, Technology, and Logistics, as part of M-code receiver card acquisition planning, assign an organization with responsibility for systematically collecting integration test data, lessons learned, and design solutions and making them available to all programs expected to integrate M-code receiver cards.

Agency Comments and Our Evaluation

We provided a draft of this chapter to the Department of Defense for review and comment. In its written comments, reproduced in appendix II, DOD concurred with the recommendation. DOD also provided technical comments, which we incorporated as appropriate.

Cristina T. Chaplain
Director, Acquisition and Sourcing Management

List of Committees

The Honorable John McCain,
Chairman

The Honorable Jack Reed,
Ranking Member
Committee on Armed Services,
United States Senate

The Honorable Thad Cochran,
Chairman

The Honorable Richard J. Durbin,
Ranking Member
Subcommittee on Defense Committee on Appropriations,
United States Senate

The Honorable Mac Thornberry,
Chairman

The Honorable Adam Smith,
Ranking Member
Committee on Armed Services House of Representatives

The Honorable Kay Granger,
Chairwoman

The Honorable Pete Visclosky,
Ranking Member,
Subcommittee on Defense House of Representatives

APPENDIX I: OBJECTIVES, SCOPE, AND METHODOLOGY

To determine the extent to which there are acquisition risks to sustaining the Global Positioning System (GPS) satellite constellation, we reviewed the Air Force GPS quarterly reports, program acquisition baselines, integrated master schedules, acquisition strategies, software development plans, test plans, and other documents to the extent they existed for GPS III, Next Generation Operational Control System (OCX), and Contingency Operations (COps) programs. We also interviewed officials from the GPS III, OCX, and COps programs; the Air Force Space and Missile Systems Center's (SMC) GPS Enterprise Integrator office; the prime contractors from all three programs; the Defense Contract Management Agency; the Office of Cost Assessment and Program Evaluation; and the Office of the Director, Operational Test and Evaluation (DOT&E). We also reviewed briefings and other documents from each to evaluate program progress in development. We assessed the status of the currently operational GPS satellite constellation, interviewing officials from the Air Force SMC GPS program office and Air Force Space Command.

To assess the risks that a delay in the acquisition and fielding of GPS III satellites could result in the GPS constellation falling below the 24 satellites required by the standard positioning service and precise positioning service performance standards, we employed a methodology very similar to the one we had used to assess constellation performance in 2009, 2010, and 2015.[1] We obtained information dated May 2016 from the Air Force predicting the reliability for 63 GPS satellites—each of the 31 current (on-orbit as of July 2017) and 32 future GPS satellites—as a function of time. Each satellite's total reliability curve defines the probability that the satellite will still be operational at a given time in the future. It is generated from the product of two reliability curves—a wear-out reliability curve defined by the cumulative normal distribution, and a random reliability curve defined by the cumulative Weibull distribution.[2] For each of the 63 satellites, we obtained the two parameters defining the cumulative normal distribution, and the two parameters defining the cumulative Weibull distribution. For each of the 32 unlaunched satellites we included in our model, we also obtained a parameter defining its probability of successful launch, and its current scheduled launch date. The 32 unlaunched satellites include 10 GPS III satellites currently under contract and 22 GPS III satellites planned for contract award in late 2018; launch of the final GPS III satellite we included in our model is scheduled for October 2031. Using this information, we generated overall reliability curves for each of the 63 GPS satellites. We discussed with Air Force and Aerospace Corporation representatives, in general terms, how each satellite's normal and Weibull parameters were calculated. However, we did not analyze any of the data used to calculate these Air Force provided parameters.

Using the reliability curves for each of the 63 GPS satellites, we developed a Monte Carlo simulation to predict the probability that at least a given number of satellites would be operational as a function of time, based on the GPS launch schedule as of May 2016. We conducted several runs of our simulation—each run consisting of 10,000 trials—and generated

---

[1] GAO, Global Positioning System: Significant Challenges in Sustaining and Upgrading Widely Used Capabilities, GAO-09-325 (Washington, D.C.: Apr. 30, 2009); Global Positioning System: Challenges in Sustaining and Upgrading Capabilities Persist, GAO-10-636 (Washington, D.C.: Sept. 15, 2010); and GPS: Actions Needed to Address Ground System Development Problems and User Equipment Production Readiness, GAO-15-657 (Washington, D.C.: Sept. 9, 2015).

[2] The Weibull distribution is a common two-parameter continuous probability distribution; it is used to model the random failures of GPS satellites.

"sawtoothed" curves depicting the probability that at least 24 satellites would still be operational as a function of time.[3] We then used our Monte Carlo simulation model to examine the effect of delays to the operational induction of the GPS III satellites into the constellation. We reran the model based on month/year delay scenarios, calculating new probabilities that at least 24 satellites would still be operational as a function of time, determining in terms of month/year the point at which a satellite would be required to enter operations to maintain an uninterrupted maintenance of the 95 percent probability of 24 satellites in operation. The Air Force satellite parameters we used for the Monte Carlo simulation pre-dated the Air Force investigation into navigation payload capacitors and the subsequent decision to launch the first satellite "as is" with questionable parts. Therefore, the reliability parameters for this satellite were not informed by any possible subsequent Air Force consideration of the decision to launch the first GPS III satellite "as is" with these parts.

To determine the extent to which the Department of Defense (DOD) faces acquisition challenges developing a new ground system to control the broadcast of a modernized GPS signal, we reviewed Air Force program plans and documentation related to cost, schedule, acquisition strategies, technology development, and major challenges to delivering M-code Early Use (MCEU) and OCX blocks 1 and 2. We interviewed officials from the MCEU and OCX program offices, SMC GPS Enterprise Integrator office, DOT&E, and the prime contractors for the two programs. For OCX, we also reviewed quarterly reviews, monthly program assessments, and slides provided by Raytheon on topics of our request. We also interviewed Office of Performance Assessments and Root Cause Analyses officials regarding root causes of the OCX program's cost and schedule baseline breach and Defense Contract Management Agency officials charged with oversight of

[3] Monte Carlo simulation refers to a computer-based analysis that uses probability distributions for key variables, selects random values from each of the distributions simultaneously, and repeats the random selection over and over. Rather than presenting a single outcome—such as the mostly likely or average scenario—Monte Carlo simulations produce a distribution of outcomes that reflect the probability distributions of modeled uncertain variables.

the OCX contractor regarding cost and schedule issues facing the program's development efforts, major program risks, and technical challenges.

To determine the extent to which DOD faces acquisition challenges developing and fielding modernized receiver cards across the department, we reviewed Air Force program plans and documentation related to M-code GPS User Equipment (MGUE) increment 1 cost, schedule, acquisition strategy, and technology development. We interviewed officials at the Air Force SMC GPS program office, MGUE program office, DOT&E, and the three MGUE increment 1 contractors— L3 Technologies, Raytheon, and Rockwell Collins. To identify the military services' respective development efforts and challenges in integrating MGUE with their lead platforms, we interviewed officials from the lead program offices for the Army's Defense Advanced GPS Receiver Distributed Device/Stryker, Air Force's B-2 aircraft, Navy's DDG-51 Arleigh Burke class destroyer, and Marine Corps Joint Light Tactical Vehicle. Additionally, to understand the extent to which DOD has a plan for implementing M-code for the warfighter, we analyzed DOD Positioning, Navigation, and Timing (PNT) plans and other DOD memorandum on GPS receiver cards. We also held discussions with and received information from officials at Office of the Undersecretary of Defense for Acquisition, Technology, and Logistics; Joint Staff / J-6 Space Branch; and military service officials from the offices responsible for developing M-code receiver card implementation plans.

We conducted this performance audit from February 2016 to December 2017 in accordance with generally accepted government auditing standards. Those standards require that we plan and perform the audit to obtain sufficient, appropriate evidence to provide a reasonable basis for our findings and conclusions based on our audit objectives. We believe that the evidence obtained provides a reasonable basis for our findings and conclusions based on our audit objectives.

Appendix II: Comments from the Department of Defense

OFFICE OF THE ASSISTANT SECRETARY OF DEFENSE
3600 DEFENSE PENTAGON
WASHINGTON, DC 20301-3000

ACQUISITION

NOV 3 0 2017

Ms. Cristina Chaplain
Director, Acquisition and Sourcing Management
U.S. Government Accountability Office
441 G Street, NW
Washington DC 20548

Dear Ms. Chaplain,

This is the Department of Defense (DoD) response to the Draft Government Accountability Office (GAO) Final Report, GAO-18-74, Global Positioning System (GPS): Better Planning and Coordination Needed to Improve Prospects for Fielding Modernized Capability, dated October 2017 (GAO Code 100643).

DoD acknowledges receipt of the draft final report and concurs as written.

My point of contact is Dr. Charles (Chuck) Frizzelle at 571-372-4408 or via email at charles.d.frizzelle.civ@mail.mil.

Sincerely,

James A. MacStravic
Acting Assistant Secretary of Defense
for Acquisition

GAO Draft Report Dated September 20, 2017
GAO-18-74 (GAO CODE 100643)

"GLOBAL POSITIONING SYSTEM: BETTER PLANNING AND COORDINATION NEEDED TO IMPROVE PROSPECTS FOR FIELDING MODERNIZED CAPABILITY"

DEPARTMENT OF DEFENSE COMMENTS TO THE GAO RECOMMENDATION

RECOMMENDATION 1: The Government Accountability Office (GAO) recommends the Under Secretary of Defense for Acquisition, Technology, and Logistics should, as part of M-code receiver card acquisition planning, assign an organization with responsibility for systematically collecting integration test data, lessons learned, and design solutions and making them available to all programs expected to integrate M-code receiver cards.

DoD RESPONSE: Concur as written.

Appendix III: GPS Modernization Cost Increases, Original Baseline vs. Current Estimate

Table 5. GPS Modernization Costs

Program	**Original Baseline[a] (fiscal year 2017 dollars, in millions)**	**Current Estimate (fiscal year 2017 dollars, in millions)**	**Percentage Increase**
GPS III[b]	4,275.2	5,834.1	36
Next Generation Operational Control System (OCX)	3,591.8	5,498.4	53
Military GPS User Equipment (MGUE) Increment 1	1,505.7	1,505.7	0
Contingency Operations (COps)	0.0	161.7	N/A
M-Code Early Use (MCEU)	0.0	119.5	N/A
Total[c]	9,372.7	13,119.4	40

Source: GAO analysis of Department of Defense data. | GAO-18-74.

[a] Dates for original baselines are: GPS III–May 2008, OCX–November 2012, and MGUE–January 2017.

[b] The original GPS III baseline was for eight satellites. The current baseline is for 10 satellites. The cost per satellite has increased from $534 million to $583 million, or approximately 9 percent.

[c] The total cost does not include the cost for the GPS Enterprise Integrator which integrates, synchronizes, tests, and verifies the GPS programs but is not a baselined program of record. The fiscal year 2018 budget request projects total GPS Enterprise Integrator expenditures across numerous fiscal years as $801.9 million. The table also omits future potential development for GPS III satellites or their procurement (up to 22 in total), the planned MGUE increment 2 program, or the costs to operationally test, integrate, and procure receiver cards across the Department of Defense. The table also does not include costs to operate and maintain any of the programs once operational.

Appendix IV: Staff Acknowledgments

In addition to Cristina T. Chaplain, David Best, Assistant Director; Jay Tallon, Assistant Director; Karen Richey, Assistant Director; Pete Anderson; Andrew Berglund; Brandon Booth; Brian Bothwell; Patrick Breiding; Erin Carson; Connor Kincaid; Jonathan Mulcare; Sean

Sannwaldt; Alyssa Weir; Robin Wilson and Marie P. Ahearn made key contributions to this chapter.

APPENDIX V: ACCESSIBLE DATA

Data Tables

Status of Weapon Systems That Have Determined the Cost Needed to Transition to M-code Receivers through Fiscal Year 2021, as of February 2017

- Military-code transition fully funded: 28
- Military-code transition partially funded: 72
- Military-code transition unfunded: 616

Figure 11 Status of Weapon Systems That Have Determined the Cost Needed to Transition to M-code Receivers through Fiscal Year 2021, as of February 2017

- Military-code transition fully funded: 28
- Military-code transition partially funded: 72
- Military-code transition unfunded: 616

Agency Comment Letter

Text of Appendix II: Comments from the Department of Defense

Nov. 30, 2017
Ms. Cristina Chaplain
Director, Acquisition and Sourcing Management

U.S. Government Accountability Office 441 G Street, NW
Washington DC 20548 Dear Ms. Chaplain,

This is the Department of Defense (DoD) response to the Draft Government Accountability Office (GAO) Final Report, GAO-18-74, Global Positioning System (OPS): Better Planning and Coordination Needed to Improve Prospects for Fielding Modernized Capability, dated October 2017 (GAO Code l00643).

DoD acknowledges receipt of the draft final report and concurs as written.

Sincerely,
James A. MacStravic
Acting Assistant Secretary of Defense for Acquisition

Department Of Defense Comments to the GAO Recommendation

Recommendation 1: The Government Accountability Office (GAO) recommends the Under Secretary of Defense for Acquisition, Technology, and Logistics should, as part of M- code receiver card acquisition planning, assign an organization with responsibility for systematically collecting integration test data, lessons learned, and design solutions and making them available to all programs expected to integrate M-code receiver cards.

DoD response: Concur as written.

INDEX

A

acquisition phase, 117
acquisitions, x, 16, 33, 70, 108, 121, 191, 200
aeronautics, vii, 1, 112, 129, 136, 142
aerospace, viii, 77, 94, 95, 97
aerospace firms, viii, 77, 94
Air Force, xi, 85, 100, 105, 163, 164, 165, 166, 167, 169, 170, 171, 172, 173, 174, 175, 176, 178, 179, 180, 181, 182, 183, 184, 185, 186, 187, 188, 189, 190, 191, 192, 193, 194, 195, 196, 197, 200, 201, 204, 205, 206, 207
air traffic control system, 101
astrophysics, x, 107, 110, 111, 117, 122
atmosphere, 16, 71, 91
atmospheric pressure, 87

B

Boeing, vii, viii, 2, 3, 5, 8, 9, 10, 12, 13, 14, 15, 16, 18, 19, 20, 21, 23, 25, 26, 27, 29, 30, 37, 40, 41, 42, 54, 55, 57, 58, 59, 60, 61, 62, 63, 64, 65, 66, 67, 68, 69, 70, 71, 72, 73, 74, 75, 92, 93, 96

C

certification, viii, 2, 3, 5, 6, 8, 10, 11, 12, 14, 15, 16, 17, 18, 20, 22, 24, 25, 26, 27, 29, 30, 31, 32, 34, 37, 38, 40, 41, 42, 43, 49, 50, 54, 55, 56, 57, 58, 60, 61, 63, 64, 65, 66, 67, 68, 70, 71, 73, 75, 86, 118, 190, 194
chain of command, 36
coding, 171, 181, 184, 188, 189, 190
commercial cargo program, ix, 5, 78, 85
commercial crew program, 1, iii, v, vii, viii, 1, 2, 3, 5, 6, 7, 8, 11, 12, 13, 14, 16, 17, 18, 19, 20, 21, 22, 24, 26, 30, 31, 32, 33, 34, 35, 36, 37, 38, 39, 40, 42, 43, 47, 48, 49, 50, 53, 54, 55, 56, 57, 59, 62, 63, 66, 67, 68, 70, 72, 73, 74, 75, 86
commercial launch industry, ix, 78, 84
commercial space infrastructure, ix, 78
communication systems, 8
competing interests, 24, 33, 36
complexity, 38, 50, 108, 114, 117, 148, 165, 200

contingency, 2, 3, 12, 13, 14, 21, 22, 37, 38, 39, 42, 48, 54, 63
contract vehicles, vii, 7, 59
cost, viii, ix, x, xi, 5, 6, 8, 32, 35, 36, 42, 53, 56, 60, 70, 78, 92, 93, 96, 103, 104, 107, 108, 109, 110, 111, 113, 114, 115, 116, 117, 118, 119, 120, 121, 122, 123, 124, 125, 126, 130, 131, 132, 133, 134, 137, 138, 139, 140, 152, 153, 154, 155, 163, 165, 172, 173, 181, 183, 186, 188, 190, 191, 192, 193, 197, 198, 200, 201, 202, 206, 207, 209
crew transportation, vii, viii, 2, 3, 5, 6, 7, 8, 9, 11, 12, 13, 27, 29, 31, 37, 41, 42, 43, 49, 54, 55, 56, 57, 59, 60, 61, 62, 63, 64, 66, 70, 71, 72, 73, 75, 93
crew transportation systems, vii, viii, 2, 3, 5, 6, 7, 8, 9, 12, 13, 41, 42, 54, 57, 59, 60, 62, 63, 64, 73, 75
cybersecurity, 166, 170, 171

D

data collection, 94, 99, 112
data processing, 89
Department of Commerce, 99, 102, 104
Department of Defense, ix, 33, 78, 79, 80, 100, 164, 165, 166, 169, 171, 172, 174, 175, 180, 182, 185, 188, 192, 194, 199, 203, 206, 209, 210, 211
Department of Homeland Security, 33
Department of Transportation, 101
disability, 4, 24, 28, 72

E

engineering, 33, 34, 97, 126, 172
equipment, ix, 32, 78, 86, 89, 90, 94, 97, 98, 101, 102, 143, 145, 167, 168, 177, 178, 198, 200
export control, ix, 78, 99, 100, 102, 103

F

Federal Communications Commission (FCC), x, 78, 100, 103, 104, 105
federal government, 31, 75, 85, 161, 201
first responders, 90, 105
fiscal year, x, 21, 107, 110, 117, 122, 134, 153, 154, 155, 165, 167, 170, 177, 178, 181, 183, 186, 187, 190, 193, 198, 199, 202, 209
flight, 11, 14, 15, 17, 18, 19, 20, 21, 23, 25, 27, 41, 62, 64, 65, 68, 71, 73, 79, 82, 87, 90, 110, 113, 119, 133, 137, 143, 148
formation, x, 110, 115, 130, 132, 135
funding, 7, 8, 59, 76, 94, 108, 115, 117, 120, 122, 123, 139, 155, 161, 179, 198
funds, 76, 138, 161, 177, 178, 179, 198

G

geolocation, ix, 78, 90
global positioning system (GPS), 1, iii, vi, xi, 90, 105, 163, 164, 165, 166, 167, 168, 169, 170, 171, 172, 173, 174, 175, 176, 177, 178, 180, 181, 182, 183, 184, 185, 186, 187, 189, 196, 198, 201, 202, 204, 205, 206, 207, 209, 211
governance, 32, 197
government policy, viii, 77
government spending, ix, 77, 80
governments, viii, 17, 77, 79, 95
GPS constellation, xi, 163, 164, 167, 170, 171, 173, 174, 175, 180, 181, 185, 204
ground stations, ix, 78

H

heat shield, 16, 71
history, 87, 109, 111, 122
homeland security, 105

House Committee on Transportation and Infrastructure, 101
House of Representatives, 4, 53, 107, 116, 139, 156, 203, 204
Hubble Space Telescope, x, 115, 130, 135, 137
human, viii, 2, 3, 6, 16, 17, 23, 25, 27, 30, 33, 37, 40, 43, 54, 57, 68, 70, 71, 79, 109, 122, 123, 124
human exploration programs, viii, 2, 6, 58
human resources, 122
human spaceflight, viii, 3, 16, 17, 23, 27, 30, 37, 43, 54, 55, 57, 70, 71, 109, 123

I

information sharing, 201
infrastructure, viii, ix, 5, 56, 77, 78, 80, 85, 88, 89, 100
integration, x, 14, 64, 70, 108, 116, 117, 118, 119, 120, 125, 129, 130, 131, 132, 133, 137, 138, 140, 141, 142, 143, 145, 146, 147, 148, 149, 151, 152, 153, 155, 169, 181, 187, 191, 193, 195, 199, 202, 211
integrity, 76, 161, 176
interference, x, 78, 168
International Space Station, v, 1, 3, 4, 22, 26, 41, 47, 49, 53, 54, 56, 75, 83, 86
international standards, 89
interoperability, 89, 176
investment, x, 12, 32, 56, 63, 69, 75, 94, 100, 107, 111, 191, 197
issues, ix, 12, 14, 16, 34, 35, 64, 78, 99, 111, 112, 115, 118, 119, 121, 123, 127, 131, 133, 134, 138, 139, 140, 141, 145, 146, 147, 149, 150, 151, 152, 154, 180, 183, 186, 187, 189, 190, 191, 195, 201, 207

L

launch vehicle, ix, 9, 10, 16, 17, 18, 60, 61, 64, 66, 70, 71, 72, 78, 79, 80, 81, 85, 86, 88, 90, 91, 92, 94, 95, 96, 98, 103, 109, 116, 118, 119, 141
laws and regulations, 99, 100
life cycle, 108, 110, 113, 114, 119, 120
low-Earth orbit, vii, 23, 33, 112

M

management, ix, x, 4, 31, 32, 36, 37, 43, 68, 78, 93, 100, 101, 107, 108, 110, 111, 115, 120, 121, 125, 126, 133, 134, 138, 139, 140, 152, 154, 155, 173, 191, 197, 201
manufacturing, ix, 14, 23, 64, 78, 81, 90, 94, 97, 98, 99, 118, 124, 138, 194
Marine Corps, 179, 196, 207
methodology, 7, 58, 92, 168, 204
military, xi, 80, 83, 84, 87, 100, 101, 102, 105, 163, 165, 166, 167, 168, 169, 170, 171, 173, 174, 179, 193, 195, 196, 197, 198, 199, 200, 201, 202, 207
modernization, 168, 169, 172, 173, 181

N

National Academy of Sciences, 122
National Aeronautics and Space Administration (NASA), v, vii, viii, ix, x, xi, 1, 2, 3, 4, 5, 6, 7, 8, 10, 11, 12, 13, 14, 15, 16, 17, 18, 19, 20, 21, 22, 23, 24, 25, 26, 27, 28, 29, 30, 31, 32, 33, 35, 36, 37, 38, 39, 40, 41, 42, 43, 44, 47, 48, 49, 50, 53, 54, 55, 56, 57, 59, 60, 61, 62, 63, 66, 67, 68, 69, 70, 71, 72, 73, 75, 77, 78, 79, 80, 85, 86, 92, 93, 94, 96, 100, 101, 107, 108, 109, 110, 111, 112, 113, 114, 115,

116, 117, 118, 119, 120, 121, 122, 123, 124, 125, 126, 129, 130, 131, 132, 133, 134, 136, 137, 138, 139, 140, 142, 144, 145, 146, 147, 149, 152, 153, 154, 155, 157
National Defense Authorization Act, xi, 102, 164, 166, 177, 197
National Oceanic and Atmospheric Administration, 99
National Oceanic and Atmospheric Administration (NOAA), 99
national policy, 197
national security, ix, 77, 79, 80, 91, 96, 100
national security agencies, ix, 77, 80
natural disaster, 90
navigation, ix, xi, 78, 89, 97, 163, 166, 170, 172, 176, 177, 185, 206, 207
navigation equipment, ix, 78, 90
next generation, 164, 166, 196

O

Office of Management and Budget, 191
officials, viii, xi, 2, 4, 6, 11, 14, 17, 19, 20, 22, 25, 26, 27, 28, 29, 31, 35, 41, 42, 43, 55, 58, 66, 70, 72, 73, 111, 112, 113, 122, 126, 130, 131, 133, 134, 143, 145, 146, 147, 148, 150, 151, 153, 154, 155, 164, 167, 182, 183, 184, 185, 187, 189, 192, 195, 196, 198, 200, 201, 204, 206, 207
operations, x, 31, 81, 82, 97, 100, 101, 105, 115, 118, 122, 130, 137, 139, 172, 189, 206
orbit, vii, 8, 9, 23, 28, 29, 33, 48, 56, 66, 79, 81, 82, 83, 84, 86, 87, 95, 96, 100, 110, 112, 118, 135, 137, 170, 175, 176, 185, 186, 187, 205
overlap, 56, 201
oversight, viii, 2, 24, 32, 34, 35, 37, 55, 73, 74, 76, 100, 101, 111, 120, 125, 133, 140, 161, 173, 197, 206

P

participants, vii, 7, 59, 79, 86
planetary missions, 147
planets, 110, 114, 135
platform, 177, 196
policy, 23, 30, 32, 76, 79, 91, 99, 114, 141, 145, 161
pollution, 96
portfolio, vii, 108, 109, 112, 114, 117, 121, 122, 125
positioning, xi, 163, 166, 170, 172, 174, 204, 205, 207
primary function, 86
private investment, 7, 59
private sector, 79, 98
probability, 3, 24, 28, 72, 125, 174, 205, 206
probability distribution, 205, 206
procurement, 92, 165, 168, 173, 194, 199, 200, 209
product design, 181
progress reports, 42
project, x, xi, 92, 108, 109, 111, 113, 114, 115, 116, 117, 118, 119, 120, 121, 123, 124, 125, 126, 130, 131, 132, 133, 134, 135, 136, 138, 139, 140, 141, 142, 143, 145, 146, 147, 148, 150, 151, 152, 153, 154, 155, 191

R

radio, 82, 88, 90, 94, 100
recommendations, iv, 2, 18, 34, 38, 39, 43, 47, 54, 76, 108, 117, 121, 127, 130, 140, 161, 173, 192, 197
regulations, x, 78, 79, 80, 99, 100

reliability, 33, 73, 76, 103, 161, 183, 205
remote sensing, 82, 90, 103
requirements, vii, 2, 3, 5, 6, 8, 10, 11, 14, 24, 26, 40, 41, 42, 43, 49, 50, 55, 57, 59, 60, 61, 62, 64, 70, 72, 74, 93, 118, 119, 120, 123, 126, 138, 194, 195, 196, 197, 202
reserves, 109, 114, 115, 116, 118, 119, 120, 121, 122, 123, 131, 134, 138, 139, 141, 147, 148, 151, 152, 154, 155
resources, 5, 56, 94, 114, 117, 122, 132, 155, 173, 174, 188, 189, 191, 201, 202
response, 39, 40, 57, 58, 115, 121, 133, 211
revenue, ix, 78, 89, 90
risk, 2, 3, 4, 6, 12, 14, 16, 17, 18, 19, 21, 22, 24, 27, 28, 29, 30, 31, 32, 34, 36, 37, 38, 39, 40, 42, 47, 48, 49, 58, 60, 63, 66, 67, 70, 71, 72, 73, 92, 100, 101, 111, 112, 119, 121, 123, 124, 125, 131, 133, 134, 137, 140, 141, 143, 145, 146, 147, 148, 150, 152, 155, 165, 180, 184, 187, 188, 190, 192, 193, 194, 198, 201
risk assessment, 21, 48, 119, 134, 140, 141, 146, 147, 150, 192
risk management, 29, 58, 133
rockets, viii, ix, 77, 78, 79, 80, 81, 85, 86, 87, 93, 94, 95, 101
Russia, viii, 3, 12, 37, 53, 54, 55, 56, 63, 68, 81, 86, 88
Russian spacecraft, viii, 53

S

safety, viii, 2, 3, 5, 6, 8, 13, 16, 18, 24, 25, 26, 27, 28, 29, 30, 31, 32, 33, 34, 35, 36, 37, 38, 39, 40, 41, 42, 43, 49, 50, 55, 57, 63, 66, 70, 71, 72, 73, 74, 75, 79, 80, 100, 101, 105
satellite service, 103, 104
satellite transmissions, x, 78, 103, 104
satellites, viii, ix, xi, 77, 78, 79, 80, 81, 82, 83, 84, 86, 88, 89, 91, 92, 95, 96, 97, 98, 99, 100, 102, 103, 104, 106, 163, 164, 166, 168, 169, 170, 171, 174, 175, 176, 177, 180, 181, 183, 185, 186, 187, 204, 205, 209
schedule delays, 13, 14, 54, 60, 63, 67, 115, 120, 125, 126, 130, 133, 139, 141, 172, 186, 190, 192, 202
science, vii, x, 68, 107, 110, 112, 115, 117, 119, 120, 122, 135, 137, 142, 158
Secretary of Defense, 167, 178, 190, 192, 197, 198, 202, 211
Senate, 4, 21, 115, 116, 139, 156, 203
services, iv, viii, ix, x, 5, 9, 10, 56, 61, 77, 78, 79, 80, 83, 85, 86, 88, 89, 90, 92, 94, 95, 97, 100, 104, 105, 126, 137, 168, 173, 174, 179, 193, 196, 197, 198, 199, 200, 201, 207
signals, 82, 88, 90, 97, 105, 166, 168, 170, 171, 174, 177
software, xi, 25, 27, 62, 98, 134, 136, 145, 151, 164, 167, 171, 172, 176, 181, 183, 184, 187, 188, 189, 190, 191, 194, 204
solar cells, 175
solar system, 110, 114, 135
Soyuz, 3, 5, 13, 21, 22, 23, 42, 48, 54, 55, 56, 63, 68, 86
Space Age, viii, 21, 42, 68, 77, 119, 129, 137, 141, 160
space environment, 15, 65, 81, 175
space exploration, vii, 1, 2, 5, 33, 54, 57, 97, 112, 124
space satellites, vii, 112
space shuttle, viii, 3, 4, 5, 33, 53, 56, 72, 86, 93, 96
space station, viii, 3, 12, 53, 54, 63
space technology, 16, 55, 70, 120
space telescope, v, x, 107, 108, 109, 110, 111, 114, 115, 116, 117, 121, 123, 126, 129, 130, 131, 132, 133, 134, 135, 136, 139, 140, 142, 145, 149, 157, 158, 160

space-related companies, ix, 78, 94
SpaceX, vii, viii, 1, 2, 3, 5, 8, 9, 10, 12, 13, 14, 15, 17, 18, 19, 20, 23, 25, 26, 27, 29, 30, 37, 40, 41, 54, 55, 57, 58, 59, 60, 61, 62, 63, 64, 65, 66, 67, 68, 69, 70, 71, 73, 74, 75, 85, 86, 91, 92, 93, 94, 118
structure, ix, 32, 33, 34, 35, 78, 87, 93, 100, 110, 114, 158, 160

T

technical comments, 59, 112, 155, 203
technical support, 7, 59
technological advancement, 100
telecommunications, 83, 97
telescope, vii, x, 82, 107, 109, 110, 111, 112, 114, 115, 116, 120, 130, 131, 132, 135, 137, 140, 141, 142, 143, 144, 145, 149, 160
test data, 164, 166, 202, 211
testing, x, 5, 10, 13, 14, 15, 16, 17, 20, 27, 57, 63, 64, 65, 66, 68, 70, 71, 73, 81, 108, 118, 119, 120, 125, 130, 131, 132, 137, 141, 143, 145, 146, 147, 152, 153, 165, 170, 171, 174, 180, 181, 182, 183, 184, 185, 186, 187, 189, 190, 194, 195, 196, 197, 201, 202
threats, 123, 166, 170, 197, 199, 202
tourism, 85, 94, 101, 103
transportation, vii, viii, 2, 3, 5, 6, 7, 8, 9, 11, 12, 13, 27, 29, 31, 37, 41, 42, 49, 54, 55, 56, 57, 59, 60, 61, 62, 63, 64, 66, 70, 71, 72, 73, 75, 92, 93, 94

U

U.S. Department of Commerce, 96, 98
U.S. Export-Import Bank, 94
U.S. policy, 23
unacceptable risk, 17, 70
United States, v, vi, viii, 1, 3, 4, 13, 23, 37, 48, 53, 56, 63, 68, 75, 79, 81, 91, 104, 107, 129, 156, 163, 203
universe, vii, x, 107, 110, 112, 114, 115, 126, 130, 132, 135

V

vehicles, vii, 7, 9, 59, 60, 79, 80, 81, 85, 86, 90, 91, 92, 95, 97, 98, 99, 103, 104, 166, 169, 178, 179, 198
venture capital, ix, 78, 94

W

workforce, 96, 98, 120, 121, 131, 133, 134, 140, 152, 153, 154